P9-CDN-263

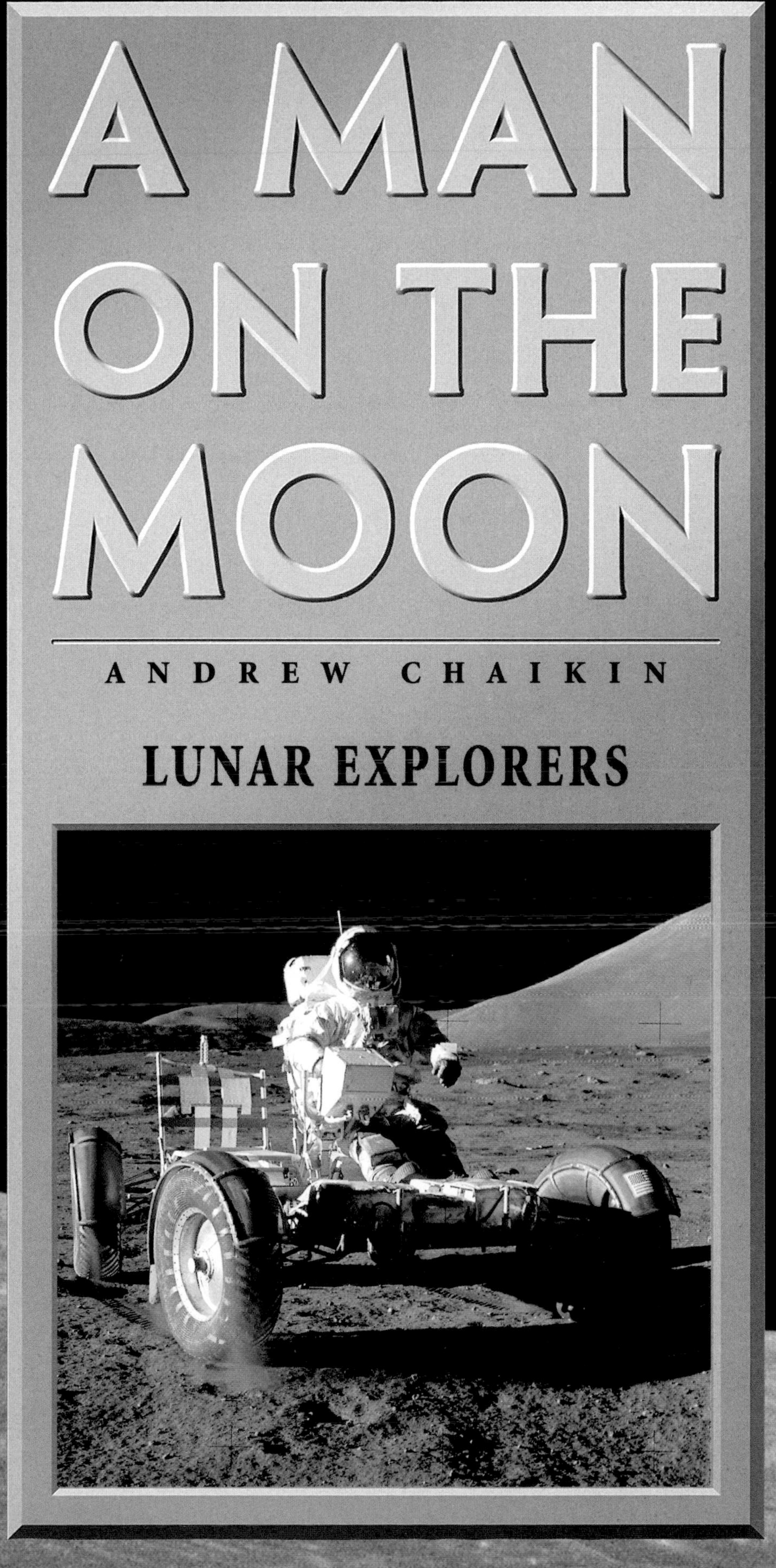
A MAN
ON THE
MOON
ANDREW CHAIKIN
LUNAR EXPLORERS

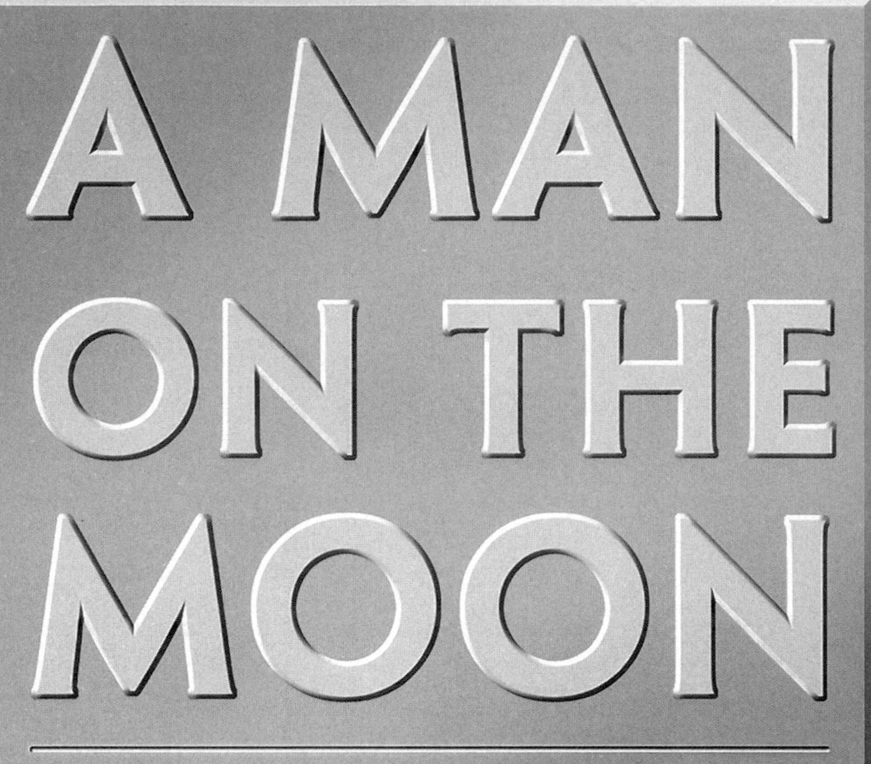

A N D R E W   C H A I K I N

III

LUNAR EXPLORERS

Commemorating
the 30th Anniversary
of the first landing on the moon,
July 20, 1969

BY ANDREW CHAIKIN AND THE EDITORS
OF TIME-LIFE BOOKS, ALEXANDRIA, VIRGINIA

THE AUTHOR

Born in 1956, Andrew Chaikin grew up in Great Neck, New York, with a fascination for the heavens and space exploration. At age 12, he made his first visit to Cape Canaveral, where he was lucky enough to meet several astronauts, including Jim Irwin *(below)*. While studying geology at Brown University, he participated in the Viking missions to Mars at the NASA/Caltech Jet Propulsion Laboratory. After graduating in 1978, he became a researcher at the Smithsonian's Center for Earth and Planetary Studies at the National Air and Space Museum in Washington. In 1980 he joined the staff of *Sky & Telescope* magazine, where he was an editor until 1986. Chaikin is now a contributing editor for *Popular Science* and has authored numerous articles for *Air & Space/Smithsonian, Discover, Popular Science, World Book Encyclopedia,* and other publications. He is a commentator for National Public Radio's Morning Edition and served as a consultant on the HBO miniseries *From the Earth to the Moon.* When he is able to take time out from writing, Chaikin pursues songwriting and performing. He lives in Arlington, Massachusetts.

CONTENTS

The first extended scientific expedition to another world, Apollo 15 was also the first moon landing to feature the Lunar Rover. Here, Apollo 15 commander Dave Scott reconnoiters near the rim of a canyon called Hadley Rille during the mission's third moonwalk in August 1971. The rille extends from the left edge of the photograph diagonally toward the horizon.

31

THE SCIENTIST

The words that John Kennedy spoke before Congress in May 1961 said nothing about going to the moon for science. There was nothing about extending the lifetime of a lunar module to three days so that astronauts could make three moonwalks instead of two, or upgrading the backpacks to nearly double the length of each trip outside. Certainly there was nothing that even hinted at developing a battery-powered car that could be folded up in the side of the lunar module like a toy in a matchbox and then, once unloaded on the lunar surface, take the astronauts miles into the distance until their lander was only a speck in the wilderness. And yet before Apollo was finished, all of this would happen, and it would happen in the name of scientific exploration. Even now, with forces already at work to bring an end to the moon program, Apollo was about to hit its stride. With these new innovations, astronauts would progress from visiting the moon to living there. And the final teams of moon voyagers, with trained eyes and hands, would visit some of the most spectacular places on the moon for the first extended scientific expeditions to another world.

Gene Shoemaker—trailblazing lunar geologist—instructs a formation of astronauts during a training expedition to Arizona's Meteor crater in April 1965. Shoemaker established the astrogeology branch of the U.S. Geological Survey in Flagstaff, Arizona.

For one man in particular, NASA's plans for the last Apollo missions were crucial. His name was Harrison Hagan Schmitt, known to friends as Jack. Dark-eyed and dark-haired, he was given to strong opinions and awful

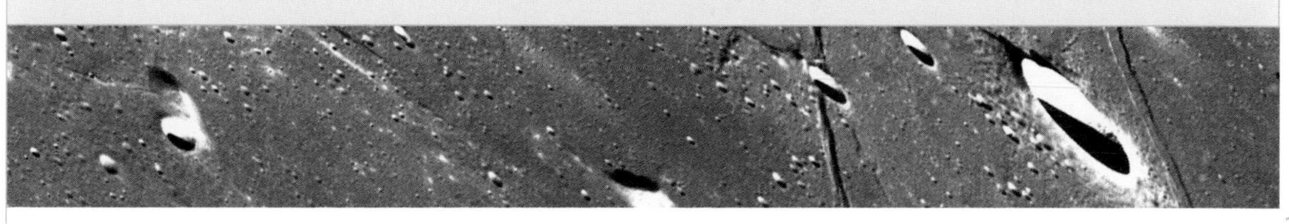

puns. He was also the first geologist-astronaut, and his ambition was to be the first scientist to practice his profession on another world. ☾

Schmitt hadn't planned to be either a geologist or an astronaut. As a teenager in Silver City, New Mexico, he spent his weekends and summers assisting his father, a respected mining geologist, in the field. When he entered Caltech in the fall of 1953 it was not to study geology but, like most Caltech undergrads, in hopes of becoming the great physicist of his generation. After half a semester as a physics major, however, he knew it would be otherwise. The next year, Schmitt moved to the more familiar ground of his father's

In early 1965, geologist Jack Schmitt tries out a lightweight aluminum frame for carrying a moonwalker's geology tools, rock sample bags, a surveying rod, and a camera. Working for Gene Shoemaker at the time, Schmitt experimented with a variety of tools and techniques for exploring the moon.

☾ Throughout this volume, a crescent at the end of a paragraph signals an author's note at the end of the book.

profession. When John Kennedy launched the Apollo program, Schmitt was a graduate student at Harvard, where his nickname was Bull Schmitt. The joke among the grad students was that of course the first man on the moon ought to be a geologist, but Schmitt showed only a casual interest in the new space program.

In spring of 1964, Schmitt was completing a postdoctoral fellowship, but the job prospects, both in his specialty of economic geology (which includes the study of ore deposits) and in academia, were dim. But some of Schmitt's friends from Caltech and Harvard had gone on to work for Eugene Shoemaker, who had established the U.S. Geological Survey's new astrogeology branch in Flagstaff, Arizona. Their reports encouraged Schmitt, and he wrote to Shoemaker, hoping there might be an opening for him.

Gene Shoemaker was one of the brilliant scientists of his generation. He'd made his mark at a young age with a study of Meteor crater, the 4,000-foot-wide hole in the Arizona desert that is one of the freshest and most recent impact craters on earth. Shoemaker made the first comprehensive geologic analysis of the crater and, in the process, reconstructed the awesome event that produced it: Fifty thousand years ago a 150-foot iron meteorite slammed into the plains and exploded with the force of a twenty-megaton nuclear bomb. No wonder the entrepreneurs and scientists who had searched for the remains of the intruder had failed: most of it had vaporized in an instant. And Meteor crater was *tiny,* insignificant compared to the hundreds of lunar craters that can be spotted with no more than a pair of binoculars. The moon's pockmarked face testifies to eons of bombardment by asteroids, left-over debris from the formation of the solar system. Shoemaker knew that the earth had endured the same terrible onslaught; anyone who wanted to see what our own planet looked like billions of years ago need only look at the moon. But the earth's ancient craters, along with most other signs of that era, have long been erased by the relentless forces of geologic activity. The surface is constantly reshaped by mountain building, volcanoes, glaciers, and erosion by wind and water. The moon, which died geologically while the earth was still young, remains a museum world. When Shoemaker looked at the moon, he saw a 4.6-billion-year history book waiting to be read.

By 1964, at age thirty-six, Shoemaker had put together from scratch, nearly single-handedly, a new scientific discipline called lunar geology. At that time the words "lunar geology" barely had any meaning to most people, but Shoemaker had made the first detailed geologic map of part of the moon, and had established a system for delineating lunar geologic time. Working from telescopic photos, Shoemaker and his USGS colleagues had staked out parcels of moonscape the way Survey geologists will tackle the corner of a state. To the Survey's map list, brimming with titles like "Geologic Map of the Sawtooth Ridge Quadrangle of Montana," they added titles such as, "Geologic Map of the Copernicus Quadrangle of the Moon." ☾

But Shoemaker's ambitions were even more far-reaching than his telescope. In 1948, as a twenty-year-old Caltech geology student, he'd read of experiments with captured V-2 rockets and foreseen the coming of space exploration; he also knew that he wanted to be part of it. Since then, he had built his career on being the first scientist on the moon. By 1962, with the

However the pilots may have felt about them, the scientists faced a very real and crucial hurdle: they had to learn to fly jets.

moon a national goal, Shoemaker's efforts went into high gear. And if NASA wasn't taking scientists into the astronaut corps, Shoemaker would do what he could to change that. But his dream was abruptly shattered in 1963 when he was diagnosed with Addison's disease. Even then, Shoemaker continued to pressure NASA to take scientists into the astronaut corps. Whoever he turned out to be, Shoemaker knew that the first geologist on the moon—the man who would go where Shoemaker so longed to go—would have an unprecedented chance for scientific discovery.

Schmitt's letter crossed in the mail with an invitation from Shoemaker to join the work under way at Flagstaff. Schmitt became part of an effort to figure out techniques for field geology on the moon. He had been at Flagstaff only a few months when, in October 1964, Shoemaker's efforts paid off: NASA announced it would select scientist-astronauts. The National Academy of Sciences would screen applicants and submit a list of recommendations to NASA. Shoemaker immediately encouraged Schmitt and two others on his staff to apply. In all, more than a thousand scientists sent their names to the Academy, far fewer than the selection committee had hoped. It appeared that most of the country's top young scientists weren't willing to gamble their careers on what they saw as a slim chance of going into space. The Academy had hoped to cull fifty or sixty names, from which NASA might select a dozen or so astronauts, but in the end they sent only sixteen names to the space agency. Three of Shoemaker's people, including Schmitt, made the list; NASA rejected all three. In fact, none of the geologists recommended by the Academy passed NASA's physical. ☾

Shoemaker was astonished. To select scientist-astronauts for a moon program and not include a single geologist was beyond his comprehension. Of all the geologists, Schmitt had come the closest to passing, and Shoemaker persevered on his behalf. None other than Randy Lovelace, the country's top aerospace physician, reviewed Schmitt's case and pronounced him fit to fly. When NASA announced the names of six new scientist-astronauts in June 1965, Schmitt was on the list.

●◐◐○○○◑◑

They were outsiders from the moment they arrived. The astronauts knew about the pressure from the National Academy—the lobbying in Washington, the letters and phone calls. And they couldn't believe Deke would have taken on a bunch of scientists if they hadn't been forced down NASA's throat. Behind their office doors, they wondered what it would accomplish. Nothing personal; but what were a bunch of scientists going to contribute? In the race to the moon, science was excess baggage.

However the pilots may have felt about them, the scientists faced a very real and crucial hurdle: they had to learn to fly jets. Two of them, physician Joe Kerwin and astronomer Curt Michel, were already qualified jet pilots. Physicist Owen Garriott had a private pilot's license, but the other three had never flown an airplane of any kind. These men—physician Duane Graveline, physicist Ed Gibson, Garriott, and Schmitt—headed to Williams Air Force Base in Arizona for a year of pilot training.

Within three weeks Graveline had left the astronaut program for personal reasons. The three who persevered at Williams, meanwhile, were soon immersed in learning to fly. After thirty hours of basic flying instruction in a Cessna 172 they moved on to jets, first the modest T-37; then, at last, the sleek, high-performance T-38. Competing against younger cadets, they graduated in the upper third of their class, with Gibson and Garriott very near the top.

But that hardly seemed to matter in the summer of 1966, as Schmitt, Gibson, and Garriott returned to the Astronaut Office with their wings, to the visible surprise of some of the pilots. In their absence, NASA had selected the Original 19, who now joined the five scientists for classroom studies. When classes ended, the scientists lost all illusions of moving ahead. The

Frank Borman *(center)* joins fellow astronauts and their geology instructors for a group portrait during a 1964 geology trip to Phantom Ranch at the bottom of the Grand Canyon. Though most of the pilots disliked classroom instruction in the subject, they happily tolerated field exercises like this one.

Nineteen were assigned to support crews; the scientists—who ostensibly had more seniority—weren't. No one had to come out and tell them what was suddenly obvious: the pecking order didn't apply to them. They were never in line. At this low point, each of the scientists faced his own private decision; each decided to tough it out.

●◐◐○○○○◑●

It was almost beyond belief to the pilots when NASA selected eleven more scientist-astronauts in the summer of 1967. Still, they had a good idea what brought it on. They knew the tension between NASA and the scientific community had not abated. They probably did not know that at NASA Headquarters, George Mueller had been pressuring Deke Slayton to get "manned up" for the Apollo Applications earth-orbit and lunar missions that would follow the Apollo landings. What Mueller didn't seem to recognize—or perhaps didn't want to believe—was that most of those missions were disappearing from the NASA budget even as the new scientists were being screened. And on the day eleven bright-eyed young scientists reported to Houston for their first day as astronauts, they didn't get the welcome they expected. Shepard's politely worded greeting boiled down to this: "We can't use you, and if you had any brains, you'd leave." They didn't even get their own offices; they were corralled into one big area that became known as Boys' Town. Once the shock wore off, the new scientists did their best to take their predicament in stride. They hung in through flight training—once more, some graduating in the top of their classes—and classroom instruction, and then went after whatever assignments were available. If they had to be the jetsam of the astronaut corps, they could at least do it with a sense of humor; they dubbed themselves the XS-11.

It was clear from the beginning that none of the XS-11 had any hope of getting a seat on a moon flight. Even Apollo Applications, whose raison d'être was science, was a long shot, since Schmitt's group was ahead of them in line for those seats. As it would turn out, the XS-11 would have to wait until the 1980s to fly in space, by which time some of them would be in their fifties. By comparison, Schmitt's group had it good.

And you would not have known Schmitt was an underdog if you had seen him in action around 1967. Slayton assigned him as the astronaut representative on the ALSEP experiments, and under his own initiative he branched out to cover the lunar module's descent stage and its cargo of gear—geology tools, television cameras, other experiments. And these were just the beginning. His bachelor apartment, across from the Manned Spacecraft Center, became a center for round-table discussions on lunar exploration. One evening Schmitt would host his geologist colleagues, the next it would be an assemblage of top-level Apollo managers. Mission strategies and goals, rock-collecting techniques, new theories of the moon's geology, and other issues were under discussion.

None of the other scientists matched Schmitt's involvement in Apollo. He was almost fanatical in his focus. If he had any life outside of Apollo—which some doubted—he never talked about it. Geologist friends would call to invite him to dinner on a Saturday night to find out he was spending the weekend working. Even when they coaxed him out for a break, he talked shop.

Shepard's politely worded greeting boiled down to this: "We can't use you, and if you had any brains, you'd leave."

There was a bar at the Nassau Bay Hotel called the Seville Club, otherwise known as the Boom Boom Room for the loud bands that played there. For one brief moment in the sixties the Boom Boom Room had topless waitresses. One afternoon a few of the geologists dragged Schmitt there for a beer, and he spent the time in lively discourse on strategies for lunar exploration. If Dick Gordon, for example, had done that, they would have checked for a pulse.

Schmitt also seemed oblivious to pressure. Aside from the scrutiny from other astronauts, there were those outside of NASA who watched him even more intently. Now that he was an astronaut, some scientists expected him to be the scientific community's inside man. But Schmitt knew his strength was that he could play both sides of the issue. To the scientific community he was a friendly representative of the astronaut corps, and he worked hard to represent the astronauts' positions and to educate the scientists in the realities of spaceflight. And to the astronauts, he was a voice of encouragement to do more science, but without the unrealistic expectations that outside scientists often brought. And increasingly, Schmitt was doing what few of his colleagues on the outside could: he was spurring the pilots' interest in geology.

That wasn't easy; most of the pilots had little or no enthusiasm for geology classes, and with good reason. Schmitt saw the problem soon after he arrived. The astronauts were being lectured to, served up a hopelessly dull plate of chemical formulae and arcane schemes for classifying rocks, the kind of material that had almost put Schmitt to sleep when he was a Caltech freshman. And when it came right down to it, geology classes were just extra work for men who already had their hands full. Schmitt saw a chance to improve things with a training program that was tailored to the specific tasks of lunar exploration. But first he would have to get Al Shepard's approval; by the fall of 1967, he had it. ☾

Neil Armstrong examines a rock sample during a geology training trip to the Quitman Mountains of west Texas in February 1969. Armstrong, who displayed more interest in geology than many astronauts did, was later praised for his great job gathering lunar samples.

A year later Schmitt, like everyone at the space center, was caught up in the tremendous push to send Apollo 8 to the moon. Frank Borman had his hands full getting ready, and asked Schmitt to put together a flight plan for the 20 hours in lunar orbit. From then on, Schmitt became Apollo 8's unofficial scientist-astronaut. When Bill Anders wanted extra preparation for his role as the first geologic observer in lunar orbit Schmitt was eager to give it. Borman never missed a chance to give Anders a jab about that, and at press conferences he would say, "Well, Bill Anders is the scientist; let him explain that." Borman didn't actually pronounce "scientist" as if it meant second-class citizen, but he didn't have to. ☾

By the time of Apollo 10, Schmitt understood that his efforts would go much farther if he could win the support of the mission commander. Thankfully, Tom Stafford was interested, and he brought his crew to several briefings with the geologists. Frank Borman's team had called the moon a black-and-white world, raising new questions about its true colors; Stafford was eager to settle the issue, and to take whatever pictures and make whatever observations the geologists wanted, as long as they didn't interfere with the mission. Not that Stafford was that interested in the moon; more than anything else, Schmitt realized, Stafford wanted his mission to stand out, and if he and his crew could settle a few lunar mysteries, it wouldn't hurt.

Unfortunately, Stafford's crew didn't have much time for science; of course, neither did Neil Armstrong's. Schmitt understood that on the first landing science had to take a back seat—a fact that Shoemaker and some of his colleagues didn't seem to understand. They protested bitterly that Apollo 11's lunar module pilot wasn't a scientist. In his most naive moments, Schmitt could agree with them—"If they were really smart, they'd put a geologist on the first landing"—despite the fact that he had barely seen the inside of a simulator. Deke Slayton obviously didn't have any such delusions; he would say, "A dead scientist on the moon wouldn't do anybody any good." A scientist on a spaceflight had to be an astronaut first and a scientist second. And in his more realistic moments, Schmitt had to agree.

More importantly, some of the scientists had underestimated the astronauts; Neil Armstrong's work on Apollo 11 made that clear. For years, the geologists had the feeling that Armstrong was genuinely interested, and that he was picking up more than most of the other pilots. They were right; Armstrong turned in an excellent performance on the moon. In the postflight debriefings he was full of detailed comments on what he had seen, and he made clear the potential for a scientific observer on the moon. ☾

With that in mind, Schmitt and other geologists—including a friend from his days at Flagstaff, Gordon Swann—eagerly began training Pete Conrad and Alan Bean for Apollo 12. By the time they left earth, Conrad and Bean were excellent geologic observers. But on the moon, to Schmitt's great surprise, they seemed to avoid talking about what they saw in detailed geologic terms. At one point Conrad mentioned a rock that had a glint of green in it, the color of a ginger ale bottle. Schmitt realized immediately that Conrad was talking about the mineral olivine—and he knew Conrad knew it too.

Schmitt was convinced that one reason for Conrad's reluctance had come from some of the geologists. They had been listening when Buzz Aldrin radioed a description of a sparkling rock at Tranquillity Base which he said looked like "some sort of biotite." The mineral biotite is a form of mica, and mica can only form where there is water. The moon has no water, and hence no mica. Behind their closed office doors these scientists complained, "Why try to teach them anything?" But these scientists missed the point. Aldrin didn't say it was biotite; he said it *looked* like biotite—he even hedged his bet by adding, "We'll leave that for later analysis." He was doing exactly as he'd been trained, describing what he saw as best he could. But these few scientists raised such an outcry that word got back to Conrad and Bean during training. After that episode, Schmitt could understand why Pete Conrad didn't want to risk making a "dumb shit" mistake on the moon, with the world listening.

But for Pete Conrad, geology was not at the top of the list. Until July 1969, he was still preparing to fly the first landing in case Apollo 11 failed;

even afterward, he was absorbed in perfecting the pinpoint landing, visiting the Surveyor, and so on. But that changed with Apollo 13. For the first time, the objective was geologic exploration. For Apollo 13, Schmitt redoubled his efforts. He knew what was needed more than anything was a professional teacher, someone who could relate to the pilots and inspire them, who could do what those dull classroom lectures could not. In the summer of 1969, Schmitt called on his friends at Harvard and Caltech for help. Gene Shoemaker suggested that Schmitt contact a man who had been one of Schmitt's professors. His name was Lee Silver. ☾

Leon T. Silver was a scientist of great standing, known not only as a skilled and thorough field geologist but a superb geochemical analyst. He had made his career roaming the desert and mountain country of the southwestern United States and trying to decipher the earliest portion of the geologic

The Orocopia Mountains rise up from the hottest desert in the country. They stand naked, devoid of vegetation, in unrelenting 100-degree heat.

record. This was the expanse of time geologists call the Precambrian, everything that happened prior to the explosion of multicellular life some 570 million years ago. The Precambrian is truly the *terra incognita* of geology. Reading the story in these rocks is difficult; often they have been crushed, or melted and re-formed, so that their original character is all but unrecognizable. Sometimes it is accomplishment enough simply to determine how old such rocks are. Geologists do that by reading one of several isotopic "clocks" contained in them, and in the late 1950s Silver made some pivotal refinements to the method, specifically for measuring the decay of uranium into lead. That work won him high honors, including membership in the National Academy of Sciences. In the 1960s Silver, with his grad student Mike Duke, turned his attention to a certain class of meteorites that they felt might very well resemble what astronauts would find on the moon. By the time the first lunar samples were on earth, Silver, at age forty-four, was one of the scientists named to study them. Like Shoemaker, he was eager to see a trained geologic observer walk on the moon. ☾

Schmitt knew Silver well. At Caltech Silver was known above all as a gifted teacher of field geology. It is hard physical work to go into the field, and after ten or fifteen years most geologists hand over their classes to someone younger, but Silver never did. He was a vigorous intruder in the wilderness, his thinning, golden red hair concealed under a floppy hat, his sunburned nose protruding from a pair of sunglasses. It was his boundless energy and enthusiasm that made the difference to his students; after one lecture they knew that they had encountered a man who would make an imprint on them. Schmitt was certain he could do the same for the astronauts.

But would the astronauts give Silver a chance? Schmitt sought out Apollo 13's Fred Haise, whom he knew to be one of the most enthusiastic geology students. With Haise's help, Schmitt arranged a meeting between Haise, Lovell, and Silver at a coffee shop in Cocoa Beach. Silver made a proposal: he would take Lovell and Haise and their backups, John Young and Charlie Duke, to a place he'd picked, for a trial run. "I'm personally convinced it's not going to be a waste of your time," he told Lovell and Haise. "But you've got to convince yourselves. I'm willing to put the time in; are you?"

It was up to Lovell; Silver could tell that he was still skeptical. "We'll give you this one trip," Lovell said, "and if it works out, we'll see about doing it again." In late September, Silver loaded a party of seven—including Lovell and Haise, Young and Duke, Schmitt, and a field assistant—into one Caltech carryall and headed for southern California's Orocopia Mountains. There was nothing official about it. They paid their expenses out of their own pockets and took the trip out of their vacation time. But in that place—a testing place, for all of them—the astronauts entered a new realm, as students with a master.

The Orocopia Mountains rise up from the hottest desert in the country. They stand naked, devoid of vegetation, in unrelenting 100-degree heat. They are a geologist's paradise, brimming with textbook examples of geologic structures. Variegated layers of rock, in shades of red, yellow, orange, white, brown, are draped and folded, angled against one another, cut by faults, twisted by ancient spasms, now frozen in a time exposure of planetary history. Arriving in this geologic wonderland, Silver led the astronauts to a little half-valley that looked out on the bare hills. Silver turned and pointed to a gnarled ironwood tree a few yards away. "That tree is the LM," he said. "You've just landed. I want you to look out the window of the lunar module and describe what you see."

One by one, Lovell, Haise, Young, and Duke tried their hand. With rough verbal sketches, they rendered the lines of hills that stood before them. Silver listened; then he coaxed them: What about the layers in that mountainside? What about the texture of that rock—how would you describe that? With impressive speed, the pilots caught on, and Silver pushed them to do even more. Soon they were proving to themselves the very thing Silver wanted most to teach them: years of test flying had already given them the skills they needed to be excellent scientific observers.

For eight days Silver led his new students through a kind of geology boot camp: up at dawn, out into the field after breakfast, working right up to din-

At their campsite in California's Orocopia Mountains, the prime and backup crews for Apollo 13 wait to be served dinner by their geology mentor, Lee Silver, who took this photograph. Clockwise from near left: Fred Haise, geologist Tom Anderson, John Young, NASA geologist John Dietrich, Jim Lovell, and Charlie Duke.

ner time. When nightfall came, Silver would lead them by flashlight back over the area they studied. They talked geology from the time they awoke until they went to bed. When the experiment was over, Lovell agreed that it had been a good experience, and that they would schedule regular field trips. Young and Duke, who were looking forward to their own explorations on Apollo 16, were sold on the whole program. Silver had four astronauts hooked. Meanwhile, Lovell and Haise emerged from the desert, grungy and unshaven, in time to attend the annual meeting of the Society of Experimental Test Pilots in Los Angeles. They walked into the lobby of the Beverly Hilton hotel looking like a couple of outlaws and approached the desk, and as the clerk stared in surprise Lovell said, "This is a stickup."

While Silver was working miracles in the California desert with commanders and lunar module pilots, other scientists fought their own battle for the command module pilots. The breakthrough came with Farouk El-Baz, a young, exuberant Egyptian-born geologist with powers of persuasion. He worked for Bellcomm, the Washington-based think tank hired by NASA Headquarters as a kind of scientific "Tiger Team," that helped tackle anything from picking out landing sites to acting as a liaison with the scientific community. One of the first things El-Baz did after joining Bellcomm in 1967 was to sift through all 4,322 pictures from the unmanned Lunar Orbiter probes and list every dome, crater, ridge, bump, and knob he could identify. From that exhaustive survey, which took three months, El-Baz formulated a list of sixteen candidate sites which together offered an example of every type of feature on the face of the moon. That effort turned El-Baz into a walking lunar data bank.

By 1970, El-Baz had already done some limited work with the lunar crews, including Tom Stafford's. But as for a real program of scientific observation, there was none. Mike Collins had little time to think about the moon when he was getting ready to go there; Dick Gordon didn't have much more. But for Apollo 13, with its emphasis on science, there seemed to be a new opportunity. El-Baz talked to space center geologist Mike McEwen, who told him he might be able to arrange a meeting with someone El-Baz had never met, Ken Mattingly.

When Mattingly heard that a geologist named Farouk El-Baz wanted to talk to him about the moon, he thought it was a joke. "*Farouk El-Baz?* Have we run out of geologists in this country?" Face to face with El-Baz, Mattingly grudgingly agreed to a briefing at the Cape. El-Baz found the conference room ahead of time and covered the walls with spectacular Lunar Orbiter

Lovell *(left)* and Haise pause during one of Silver's field exercises in the Orocopia Mountains in September 1969. Though destined never to reach the moon's surface, the pair set a high standard for others to emulate, as they learned how to explore the moon.

panoramas, marked to show the orbital path of Apollo 13. When Mattingly arrived, with Lovell and Haise, El-Baz got to the point. "Anyone can look," he told them, "but few really see. We don't know very much about the moon. You have a chance to help us know more." Then he took the men on a tour of the moon, reciting from his mental data bank of landmarks and lunar mysteries, describing the lay of the land and sneaking in a good bit of geology along the way. He could tell Lovell and Mattingly were still apprehensive, but Haise was smiling with enthusiasm. Then they asked questions, and they listened to the answers. They began to make up nicknames for funny-looking craters they would use as landmarks. The meeting was supposed to go two hours; it went five. When it ended, Lovell agreed to schedule repeats. In time, El-Baz was meeting with Mattingly regularly, preparing him for a solo mission of scientific observation.

And so the command module pilots had their scientific mentor. Like Lee Silver, El-Baz had the infectious enthusiasm necessary to spark curiosity in the astronauts. And he had independently made the same discovery Silver had: once the astronauts called science their mission, their competitive, perfectionist energy did the rest. Some time later, as El-Baz stood in the lobby of Building 4 at the Manned Spacecraft Center, a thin, red-headed man with a boyish face approached him. "Hey, you're Farouk El-Baz, aren't you? I'm Stu Roosa. I want you to make me smarter than Ken Mattingly!"

Had fate not intervened in the fortunes of Apollo 13's newly trained scientific observers, Lovell, Mattingly, and Haise would probably have set a new standard for the astronauts who followed. The geologists sweated out Apollo 13's return along with everyone else, but when it was all over, they mourned the mission they had lost. Unfortunately, the prospects for Apollo 14 weren't so bright. Alan Shepard and Ed Mitchell were among the smartest men in the Astronaut Office. There was no doubt that if they had really wanted to, they could have met or even exceeded Lovell and Haise's performance. But that wasn't what Shepard had in mind. One NASA geologist who was assigned to train Big Al was warned by his colleagues, "Don't be alarmed when he acts like this isn't very serious." During briefings, Shepard would tell jokes to whoever was sitting next to him, including his lunar module pilot. Whatever Mitchell's enthusiasm, it was apparently so dampened by Shepard's disinterest that another geologist later termed Apollo 14 "the nadir of my efforts." Some believed that if Shepard and Mitchell had gone in more prepared, they would have found Cone crater.

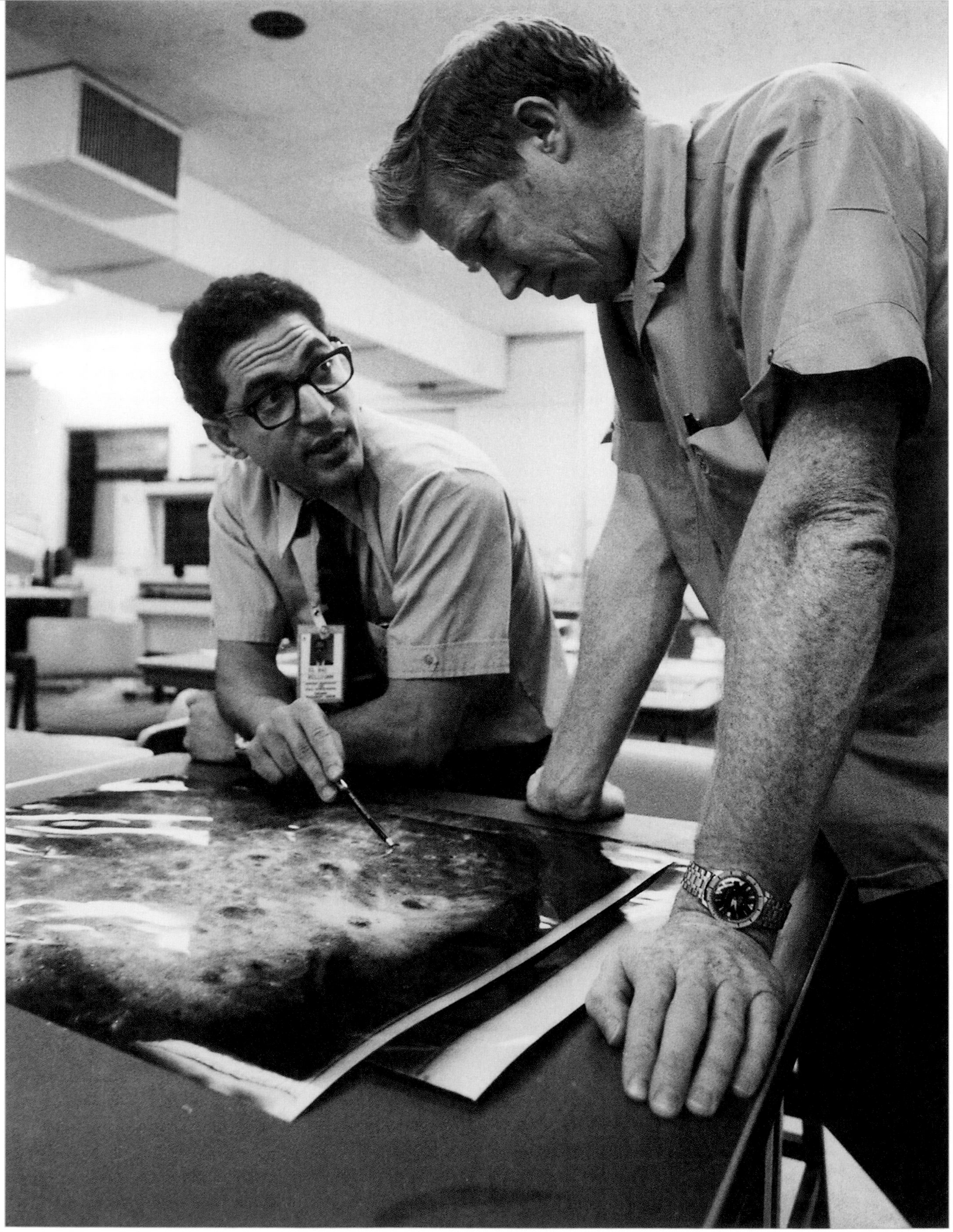

Farouk El-Baz *(left)*, seen here with Stu Roosa, coached command module pilots for their roles as orbiting explorers. The Egyptian-born geologist worked for the Washington-based company Bellcomm, which assisted in the scientific planning for the lunar landings.

Looking back, Schmitt would allow as how Shepard had other things on his mind, like getting up to speed after a ten-year absence from spaceflight, and the fact that he was about to fly the first flight after a near-disaster. Whether his attitude would have been any different under other circumstances, no one could say. In any case, the geologists learned a lesson with Apollo 14: the commander sets the tone.

All this time, Jack Schmitt waged his own personal battle to win a seat on a lunar mission. Of course, as the only geologist, Schmitt knew he had an edge over the other scientists; if any of them were going to the moon, it would be himself. And his work with the lunar crews had won him greater standing in the Astronaut Office. Schmitt's biggest worry was that Shepard or Slayton might disqualify him on technical grounds. His edge as a geologist would mean nothing if he didn't measure up as an astronaut. And so Schmitt became a simulator hound. And when Fred Haise needed a warm body to help with the Crew Compartment Fit and Function tests at Grumman, Schmitt was happy to volunteer, again and again. By December 1969, he was feeling more confident that if he got the chance, he would not be found wanting. The newspapers were reporting that he would soon be named to a crew. Soon after the new year began, Shepard called Schmitt into his office and told him the same thing. He added that Schmitt should begin stealing simulator time, unaware that he had been doing that all along. By March 1970, it was official: Jack Schmitt was the backup lunar module pilot for Apollo 15. Now, working as an insider, Schmitt would help spark a competitive energy among the prime and backup teams of astronauts for the mission that would at last begin to realize Apollo's potential for scientific exploration. ☾

Standing on a field of volcanic rock that was once a broiling lava flow, Lovell and Haise participate in a geology field trip to Hawaii's Mauna Kea early in 1970. For the astronauts, training for their scientific roles meant discovering some of earth's most spectacular landscapes.

A FIRE TO BE LIGHTED

MAY 1970

On the highway that leads from Los Angeles to the Sonoran desert, a mud-splattered truck many field seasons beyond ever being clean again sped east with Lee Silver at the wheel, and a couple of rented four-wheel-drive vehicles following behind. Two hours from Pasadena, they pulled off the road into the Orocopia Mountains. Silver led his party on foot to a small valley strewn with cobbles and boulders. When winter rains were especially heavy, this dry wash became the mouth of a river several miles long and a hundred yards wide. These rocks were the legacy of a hundred thousand rainy winters, rolled and nudged from the surrounding hills, an inch or two at a time.

"Get me the suite," Silver said quickly. "You've got ten minutes." Silver watched as Dave Scott, Jim Irwin, Dick Gordon, and Jack Schmitt hunted among the boulders. The challenge for Silver's new students was to capture the variety of this place, from the typical to the exotic, in about a dozen hand-sized specimens. In geologic terms, such an assortment is called a *suite* of rocks. But in this case, Silver had asked the impossible. Dave Scott would always remember this as an exercise in frustration. There was too

High in Colorado's San Juan Mountains, Lee Silver instructs the Apollo 15 prime and backup crews on a field trip in July 1970. From left: Dick Gordon, Jim Irwin, Jack Schmitt, and Dave Scott.

Apollo 15's commander, space veteran Dave Scott, had flown on the Gemini 8 and Apollo 9 missions. The test pilot's enthusiasm for geology blossomed during his training with Apollo's lunar scientists.

much variety here to be captured in just twelve samples, in just ten minutes. But that was the point; it wouldn't be any easier on the moon. In the spring of 1970, Silver was determined that picking up a suite of rocks would become as second-nature to these astronauts as flying the lunar module. And he had no doubts that with Dave Scott, his efforts would pay great dividends.

Even in a pack of overachievers like the astronaut corps, David Randolph Scott stood out. He seemed to have come straight from Central Cast ing, a six-footer with all-American good looks and built like a decathlon champion. In some circles there was a joke that if NASA ever came out with an astronaut recruiting poster, Dave Scott should be on it. Born on Randolph Air Force Base (hence his middle name), Scott was a general's son who graduated near the top of his class at West Point, then became an air force fighter pilot. He projected an admirable combination of charm and enthusiasm; he was also an able salesman. Other astronauts knew Scott as serious and businesslike, and something of a straight arrow, but even those who didn't get along with him put him at the top.

It was no accident that Scott was the first of the Fourteen to fly in space, but that experience, as copilot on Gemini 8's aborted tumble in earth orbit, left him with an albatross around his neck. Scott had lost the chance to make a two-hour space walk, and with it the chance to make his mark. Not until March 1969, when he turned in a flawless performance as Jim McDivitt's command module pilot on Apollo 9, did Scott feel he had shown the world what he could do.

Early on, Scott had shown more enthusiasm for geology than most of the pilots. He'd long harbored an interest in archaeology; as a fighter pilot stationed in Tripoli he'd visited the ruins of Roman cities in the Libyan desert. When the astronauts hiked into the Grand Canyon on one of their first field trips, Scott saw nearly two billion years of history written in twisted metamorphic rocks and perfectly exposed strata of limestone, sandstone, and shale. For the first time he understood what it meant to talk about *geologic* time, in which millennia are reduced to moments. And each outing brought new spectaculars. When Scott fell in behind Pete Conrad for geology training on Apollo 12, he wasn't there just because it was part of the mission. At his home in Nassau Bay, Scott proudly displayed his rock collection in a specially made wooden cabinet. And when Jack Schmitt suggested that he consider a program of geology training with Lee Silver, Scott needed little

Even in a pack of overachievers like the astronaut corps, David Randolph Scott stood out.

convincing. He and Irwin met with Silver the day Apollo 13 was launched, and a month later, in May 1970, Silver returned to the Orocopias with his new students. Once again, it was a testing period for both sides, but when it was over Scott gave his okay. For the first time, geology was fully integrated into the official training plan.

By the time Lee Silver got to him, Scott was already poised to give his all for science. In the first days of training for Apollo 15, he told his crew, command module pilot Al Worden and lunar module pilot Jim Irwin, that their goal was to come back from the moon with the maximum amount of scientific data possible. More than any mission commander before him, Scott could afford to make that pledge. He'd come off Apollo 12 with the gut-level knowledge that he could land a lunar module. Even earlier, Apollo 9 had made him a rendezvous expert. He'd formed close working relationships with the flight controllers and knew what the conglomerate brain of mission control could do in a crisis. He didn't sweat free-return trajectories or reentry targeting. Like every Apollo commander, he wanted his mission to stand out, and he would do that with scientific achievement. But if Scott was a man who strove hard to make his mark, he was also ready to take Lee Silver as his mentor. Silver made him want to learn, to understand what the rocks could tell. He was hooked.

●◐◐○○○○◑◑

Scott's backup, Dick Gordon, was equally enthusiastic, but not for the same reasons. He couldn't profess any innate love of geology, though he liked Lee Silver immensely. For Gordon, the only thing that mattered was crossing those last 69 miles, commanding his own crew, flying his own lunar module to a touchdown, and leaving his footprints in lunar dust. And if geology was part of that bargain, then that was fine with him.

Gordon was happy to have Schmitt as his lunar module pilot—once he was sure Schmitt wouldn't get him killed. That worry faded quickly enough. Schmitt was by no means the best pilot Gordon had ever flown with, but he was certainly adequate. If Schmitt had trouble with anything, it was with the commander-subordinate relationship; that was something of an adjustment for the strong-willed geologist who tended to pursue his own ideas without submitting them to his commander. Gordon was ideally suited to the task of keeping Schmitt under control, and over time, the test pilot and the scientist became a team. Their rapport was real, in the air and at the field site. Gordon complemented Schmitt's exhaustive eye for detail with a talent for seeing the broad scope of things. If Jack Schmitt would reach the moon on Apollo 18, then he would have an excellent field partner. The geologists could hardly wait.

Schmitt proved himself too in the simulator, not only as a systems man, but by learning to make a successful lunar landing. One day the instructors decided to let him show it. During a descent they failed Gordon's controls, but instead of letting Schmitt take over, Gordon waved his lunar module pilot aside and landed with his controls while Schmitt laughed. After more than six years, Dick Gordon wasn't about to let someone else land his spacecraft on the moon—even in the simulator. After the run was over he emerged from the simulator and stormed past the instructors' consoles.

"Tried to get me to let Jack land, didn't you?" He kept walking. "It didn't work, did it? It never will." And he was out of the room.

●◐◐○○○○◑◑

The unpleasant possibility that had hung over Dick Gordon since he returned from Apollo 12 finally came to pass in the summer of 1970 when Apollo 18 was canceled. The news hit Gordon hard. He didn't say much, but when he did, once or twice, Lee Silver tried to commiserate. And being an optimist, he didn't stay down for long. As for Jack Schmitt, he seemed to be too absorbed in the training itself to be crestfallen. But then, if Schmitt felt disappointment, he wasn't the kind to show it. Gordon, Schmitt, and command module pilot Vance Brand agreed that they would not give up without a fight. It was a long shot, but perhaps Slayton would assign them to Apollo 17. In the meantime, they would work to be the best crew around, so that Slayton's decision would be an easy one.

At the same time, the cancellation reshaped the fortunes of Dave Scott and his crew. Until that time, Apollo 15 had been slated as the last of the so-called H-missions—the limited exploration missions in which astronauts stayed no more than 33 hours on the moon, traveled on foot, and took two moonwalks. But NASA, anxious to increase the scientific yield of the final three landings, redesignated Apollo 15 as the first of the so-called J-missions. In the works well before Apollo 11, the J-missions were designed to push Apollo's capabilities to the limit. They would feature an upgraded lunar module, allowing three-day stays on the surface, and a long-duration backpack that would extend each moonwalk to as much as seven hours. The J-mission crews would explore places of greater geologic complexity. And they would make the most extensive lunar traverses yet, thanks to a battery-powered car called the Lunar Roving Vehicle—or, as it was known, the Rover. Scott could not have been more delighted. ☾

It remained to choose a destination for Apollo 15, and in mid-September the Site Selection Board convened in Houston. It had never been easy for the scientists to agree on landing sites, and this time was no exception. The various subcommittees and working groups had considered several alternatives, and had narrowed the choices to two. One group favored Marius Hills, a collection of low, dome-shaped rises that scientists believed might be small, young volcanoes. The other choice, a place called Hadley, lay along the shore of Mare Imbrium, the Sea of Rains. ☾

Like most lunar *maria,* Imbrium's lava plains sit within a giant crater called an impact basin, ringed by mountains. Along Mare Imbrium's shore lies one of the moon's great ranges, called the Apennines, whose peaks tower as much as 3 miles above the *mare.* The geologists suspected that these mountains were blocks of a primordial crust that had been thrust upward by the tremendous force of the Imbrium impact. The chunks of basalt from Tranquillity Base and the Ocean of Storms had taken geologists back to the era of *mare* volcanism. The Apennines promised to open a window on an even earlier time, perhaps all the way back to the moon's birth.

Putting a lunar module down among the Apennines was out of the question, but the lava plains that formed the valley floor were a ready-made landing strip. Judging from the topographic maps, it should be possible for a team of the astronauts to drive partway up the side of an 11,000-foot moun-

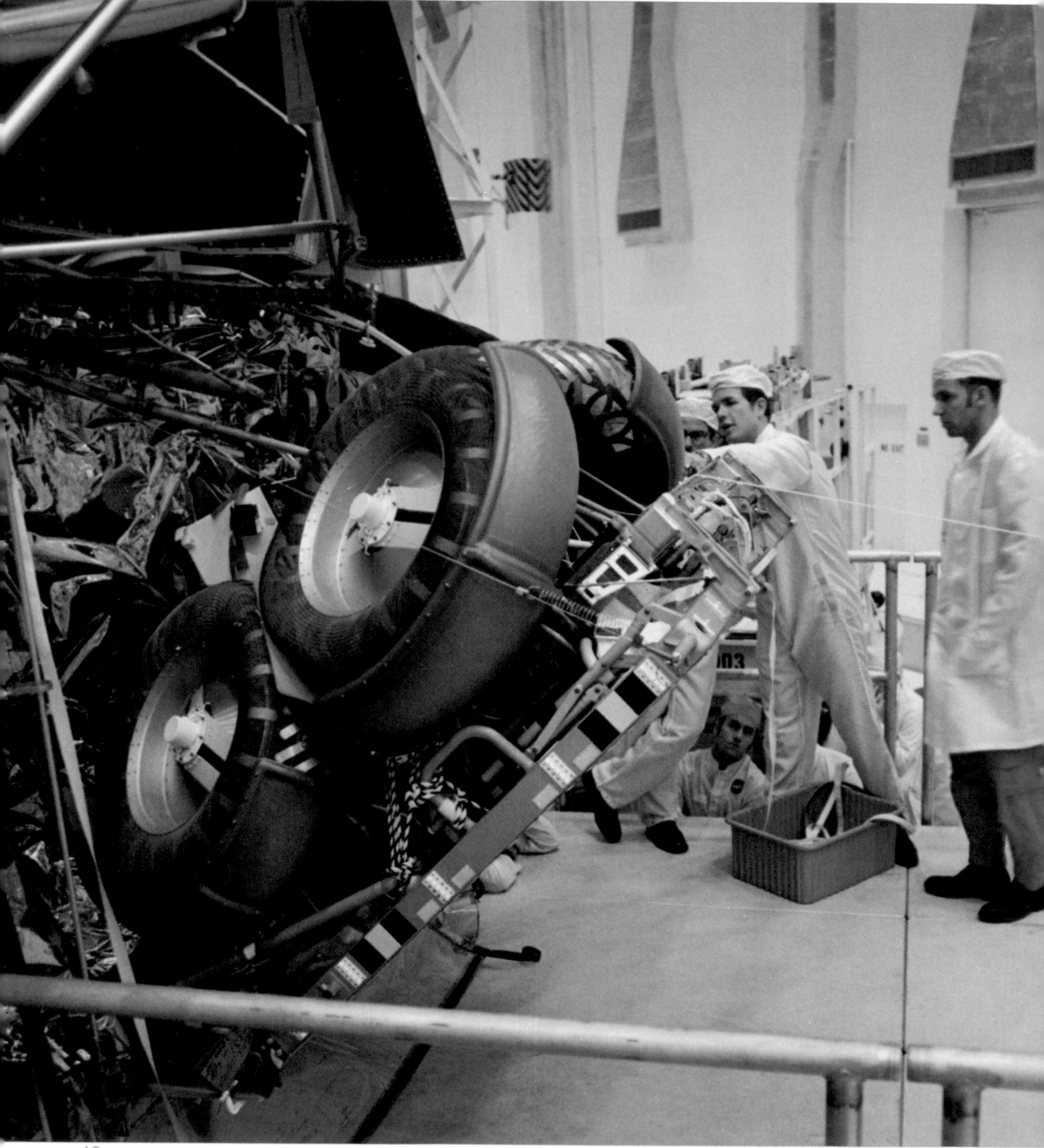

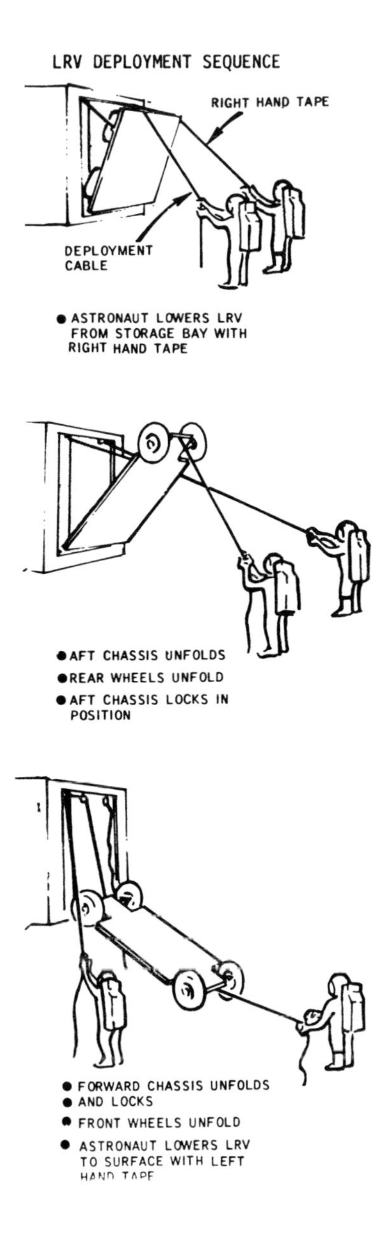

At left, Dave Scott leans forward to inspect Apollo 15's Lunar Roving Vehicle (LRV) at the Kennedy Space Center a few months before launch. The Rover's wheels are folded against the chassis in preparation for stowage in the lunar module. A sequence of NASA drawings *(above)* shows how Scott and Jim Irwin would deploy the Rover after landing on the moon.

tain called Hadley Delta and hunt for samples; a single piece of primordial crust would justify the mission. The second lure, and the one that gave the valley its name, was a sinuous channel called Hadley Rille, a mile wide and thirteen hundred feet deep; it was one of a dozen such lunar features. Many geologists suspected it had once been a river of a hellish sort, brimming with broiling lava. The first visitors to a lunar rille might be able to solve the mystery. And there were enough other attractions at Hadley—unusual craters, for example, that looked to be volcanic—to make it a geologic bonanza. ☾

The scientists were deadlocked. Although the final decision rested with Apollo program director Rocco Petrone, there was an unwritten rule that the choice had to have a thumbs-up from the mission commander. Dave Scott was in the room, and to him, the choice was easy—not only for science, but for something else he saw as essential: grandeur. It was good for the human spirit, Scott believed, to explore beautiful places. He told the scientists that although he felt confident he could land at either site, he strongly preferred Hadley; his words tipped the balance. The small valley at the foot of the Apennines became the target for the expedition.

●◐◐○○○◑●

For as long as he could remember, the mountains had been a part of Jim Irwin. Long before he knew the grand solitude that belongs to the jet pilot, he had found peace and exhilaration in high country. Some of the happiest times of his childhood had been while exploring the hills around Pittsburgh and the Wasatch mountains near Salt Lake City. As a young man in Colorado Springs he learned every hidden trail up Pikes Peak. On vacations the Irwin family scaled the rounded granite towers of Yosemite and reached the lofty summit of Mount Whitney. Before he soared to even greater heights, Irwin had stood atop many of the major peaks in the country. In the astronaut years, the mountains had been his refuge. When Houston's unbroken flatness and brown water oppressed the eye, when the steamy Gulf air was all but unbreathable, he and his wife, Mary, would escape with the children to the Rockies, to hike in the clean, rarefied air. How happy Irwin was, then, to discover that he would fly the first mission to the lunar mountains. Now age forty, Irwin was one of the oldest members of the Original 19 (and two years Scott's senior), and there was a quiet self-confidence about him, a maturity, that set him apart; perhaps almost losing his life and livelihood had reordered Irwin's sense of what mattered. In 1961 he shattered both of his legs in an airplane crash while teaching a student to fly. For a while the doctors feared he might lose a leg to gangrene. But Irwin fought his way back to flying and built himself into a figure of physical strength. He was probably among the more religious men in the Astronaut Office, but most of his colleagues would not have known that about him. He was as amiable as any of them, but he did not call any of them close friends. His best friends were his wife and children. For a test pilot, he was decidedly unflamboyant. No lunar module pilot before him had so completely avoided competing with his commander. In his own way, Irwin seemed ideally matched to Scott. Where Scott could be rigid, Irwin was deferential. When he did disagree, it was always with tact. Years later, Scott would say that Jim Irwin was probably the only person who could have gone to the moon with him; only Irwin could have put up with his authoritarian style of command. In truth, Irwin admired his commander's intelligence and energy. He was sure that if they were to send Dave Scott off to the moon all by himself that most of the mission would be accomplished. There was no question about it; Apollo 15 was Dave Scott's mission. ☾

NOVEMBER 1970
SAN GABRIEL MOUNTAINS, CALIFORNIA

The rocks were white; they lay in a jumble of slabs and chunks on the floor of a tiny canyon. They had fallen here from much higher up; Scott could see that now. Silver explained that the San Gabriels had pushed up in a spasm of mountain building 20 million years ago. The upheaval had been rapid, in geologic terms, lasting perhaps only a few million years, and even as the rocks were pushed upward they were fractured by the tremendous stresses of the event. Much more recently, these white rocks had slid off the face of the mountains, whole layers like flecks of peeling paint, into this tiny canyon.

According to the geologists' best guess, this chunky white rock, called anorthosite, was the stuff of the moon's primordial crust. One of the unmanned Surveyor landers had radioed back data suggesting the existence of this rock in the highlands near the crater Tycho. And tiny pieces of anorthosite had turned up among the coal-dark dust from Tranquillity Base and the Ocean of Storms; because of their different composition, it was obvious to the geologists that these misfits had been hurled there from some distant place—namely the highlands (including the Apennine mountains) that lay beyond the shores of the *maria*. One look at the full moon, and the contrast between the dark gray *maria* and the white highlands, spoke to the sense of the idea. ☾

Explaining why most of the moon's crust should be composed of anorthosite led some geologists to an extraordinary scenario. Within the infant satellite, they proposed, there was so much heat that the entire outer shell became an ocean of molten rock. As this "magma ocean" cooled, minerals crystallized. The heavier species, including the iron- and magnesium-rich crystals, sank to the bottom. The lighter crystals, specifically, the mineral plagioclase, which is the main component of anorthosite, floated to the top. In bringing Scott and Irwin to the San Gabriels, Silver's hope was that if they found anorthosite in the Apennines, they would recognize it from having seen these rocks.

It didn't matter how much Scott and Irwin understood about the San Gabriels, or even about the moon's magma ocean—though they seemed to take in just about everything Silver told them. What mattered, on this trip and the others that had become a regular part of the training schedule, was their ability to describe accurately what was in front of them. At first glance, the white blocks seemed to be a random jumble. But the more Scott and Irwin looked, the more they could see order—or *organization,* as Silver liked to say. Now they could see a band of darker rock running through the fragments, broken but still coherent, like the pattern on a shattered dinner plate. This was the kind of detail that would speak volumes, transmitted across the earth-moon airwaves. To Pete Conrad—"Why talk about the rocks when they'll have the rest of their lives to study them?"—Silver would have answered with a single word: context. To a field geologist, as much as a sociologist, context is crucial. Rocks, like people, are a lot easier to understand when you know where they came from and who they grew up with.

To be sure, when it came to context, the moon didn't make things easy. On the moon, small rocks are hot potatoes. They begin life as high-velocity projectiles fleeing the blast of a meteorite impact. Once they come to rest, perhaps many miles from their source, they're then likely to be bounced around by later impacts. Over the eons, a cobble-sized specimen might be kicked and nudged hundreds of miles away from the place where it formed. But the bigger a rock is, the harder it is to move. The larger the boulder, the more likely to have formed near the spot where it is found, simply because any jolt strong enough to move it would probably have broken it into pieces. As for an outcrop of true bedrock, that was something no astronaut had yet identified. The promise of outcrops, up on the slopes of the mountains, and perhaps at the rille, was one of Hadley's lures.

All the geologists were impressed with how fast the astronauts caught on to the intricacies of lunar field geology. What amazed Silver was that Scott could find time for geology at all. It would one day occur to Lee Silver that the massive effort mobilized for Apollo 15 was like a military invasion. Bob Gilruth, Chris Kraft, Jim McDivitt, and Rocco Petrone were the generals, each

Irwin *(left)* and Scott practice their sample-gathering tasks during a practice moonwalk at New Mexico's Rio Grande Gorge in March 1971. Their backpacks were lightened to simulate their weight in moon gravity.

Scott and Irwin practice driving the Grover, an earthbound rover, near a simulated lunar crater excavated with explosives in Flagstaff, Arizona. The Grover was built by the U.S. Geological Survey, which also created the mock crater field for astronaut training.

with his own theater of operations, and troops of engineers, flight controllers, crew systems specialists, and of course, contractors around the country, numbering in the tens of thousands. Silver was a captain somewhere in the infantry. There were enough changes and additions to the hardware and the mission design to make Apollo 15 seem almost like a new program. On Long Island, the Grumman people had outfitted Scott's lunar module with improvements such as bigger fuel tanks, extra batteries, and even a bigger exhaust nozzle for the descent engine, to give the LM extra thrust needed to carry the Rover and other gear to the surface. At the Boeing Company in Seattle, engineers wrestled with the fledging Lunar Roving Vehicle. Meanwhile, in Houston, trajectory specialists hammered out the steeper descent path required for Scott's lunar module to come in over the Apennine mountains.

For all of this and more, Dave Scott was the point man. All of these developments competed for his attention. And yet, Scott never turned away the scientists' concerns. He was their advocate with Petrone and the other generals in this invasion. The geologists could hardly believe the resistance that came up whenever they mentioned a new piece of gear. They wanted Scott and Irwin to carry a telephoto lens to take pictures of the mountains and the rille; Petrone and McDivitt fought it. Silver devised a rake to sift small rock samples from the soil; they fought that too. It was understandable; every pound in that lunar module was worth precious seconds of hover time; it was a matter of safety. But here again, Scott was the geologists' best ally. He would reach into his shirt pocket and pull out an index card already brimming with "action items" and add one more, and he'd knit his brow with a kind of mock concern and say to Silver, "We'll work it, Professor." ☾

Most of all, Scott made time for the field trips. The training had intensified since the spring. By summer, the trips were no longer merely teaching exercises; in his drive to get ready, Scott pushed Silver to make them more like true simulations, and the men wore backpacks, cameras, and radios. By November, a training version of the Rover was added. Everything they did was like an actual mission. They followed traverses designed by the geologists. They used aerial photomaps that had been blurred until they matched the level of detail of the best photos of Hadley. And they solved real geologic

At the Manned Spacecraft Center, Scott *(right)* guides a training version of the Rover over a sandy and rock-strewn man-made handling course, complete with wood ramps to mimic rugged terrain. The vehicle is suspended from huge shock absorbers to simulate the moon's lower gravity.

problems. The best measure of progress was the increasing sophistication of their geologic descriptions. Up on a hill, Joe Allen, the young scientist-astronaut who would serve as Capcom for the moonwalks, sat in a tent with a two-way radio. With him were geologists who were unfamiliar with the area, so that they depended on the astronauts' words. ☾

Through the end of 1970 and into 1971, Silver and his students were taking a trip every month, sometimes two, and the dynamic that was shaping up was really impressive. The healthy competition that had first showed itself in the Orocopias with the Apollo 13 prime and backup crews was in full flower. Jack Schmitt was very much a part of that chemistry. By his very presence, he raised the level of discussion. On field trips, during the review sessions, Schmitt would notice a key detail that Scott and Irwin had missed. It was easy to see that Dick Gordon was benefiting from having Schmitt as a field partner, and they in turn gave Scott and Irwin something to shoot for. There wasn't any question that each was trying to outdo the other.

Silver and his students were taking a trip every month, sometimes two, and the dynamic that was shaping up was really impressive.

Field geology is as much an art as a science, altogether different from the rigid, structured world of the simulator. Scott liked the change. He liked the hard work of the field trips, and he liked the camaraderie. Up to now Scott had not known Jack Schmitt well, but he could honestly say that he had never resented his presence, or any of the scientist-astronauts', in the Astronaut Office. The program needed the diversity they brought—but it was also clear to Scott that letting them fly before the lunar landing had been mastered was out of the question. Now that Schmitt was a key player in Apollo 15, Scott was glad for his help, though he had little patience for Schmitt's jokes—or anything from anyone that felt like a waste of time. And years later, Scott would deny that he was competing with his backup crew. He would say that his motivation was simple: He wanted to please his Professor. But that was costing him. More than once on these trips, Silver heard Scott mention how little time he had with his children. His wife, Lurton, had signed up for an introductory geology class at the University of Houston, not only to help fill

Muscovites gather in Red Square to mourn the death of Soviet cosmonauts Viktor Patsayev, Vladislav Volkov, and Georgy Dobrovolsky, who spent a record-breaking three weeks aboard the Salyut 1 space station in June 1971. The trio perished when their Soyuz 11 spacecraft suddenly lost cabin pressure before its automatically controlled reentry.

the time when Dave was away, but so that when he was home, she would have something to talk about.

By March, the team had jelled, and people from all levels of NASA were witnessing it. That month Scott convinced Gerry Griffin, who would be the flight director during the moonwalks, to join a trip to New Mexico's Rio Grande Gorge, so he would understand what Scott and Irwin would try to accomplish on the moon. In April, when they ventured into California's Coso Hills, there must have been thirty or forty people, including Rocco Petrone and other members of the NASA brass, who had come along to see just what Lee Silver and his students were up to. What they saw was an expeditionary force for a geologic assault on Hadley.

On the last day of June 1971, the space center was rocked by a reminder that even now, there was nothing routine about sending humans into space. Three Soviet cosmonauts were returning to earth aboard their Soyuz 11 spacecraft after spending twenty-four days in orbit aboard the Salyut 1 space station, shattering all space endurance records. The flight was nothing less than a milestone toward missions to the planets; Americans would not better the feat for another two years. Everything was fine as the Soyuz hurtled toward reentry under automatic control. Minutes later, recovery helicopters in Kazakhstan spotted the craft descending under an orange-and-white parachute, but were unable to contact the men by radio. After touchdown, ground crews excitedly hurried to the ship and opened the hatch. To their horror, they found the men dead, still strapped into their seats. Later, they determined that the cosmonauts had perished from a sudden air leak just before reentry.

In Houston, the deaths sent a ripple of concern through the space center. With less than a month to go before launch, managers debated changing the flight plan to have Scott's crew don their space suits for reentry. Scott vetoed the idea; just because the Soyuz had a problem didn't mean that Apollo was suddenly suspect; his faith in his spacecraft was unshaken. Meanwhile, outside NASA's world, the question that had dogged the moon program for two years still lingered: Why send people back to the moon at all? The previous fall, the Soviets had scored two landings with unmanned probes. In September, Luna 16 had brought back a sample of lunar soil. Then, in November, a wheeled robot called Lunakhod 1 was dispatched to roam the plains near the moon's Bay of Rainbows, several hundred miles from where Scott and Irwin were slated to land. Gene Shoemaker was now saying it would be better if NASA spent its money on unmanned missions, if they insisted on not sending scientists to the moon. In Scott's mind, the argument was misguided. He and Worden and Irwin would go to the moon with trained eyes and trained minds. No robot could do that. Lunakhod couldn't see what they would see, speak the words they would say, feel what they would feel, or come back to tell about it. Hadley was waiting for him and Irwin, and they would meet it with the human's capacities to probe, to take advantage of the unexpected, to make discoveries. This wasn't about technology, not anymore. It was the curiosity that now burned inside him, which Lee Silver had helped to ignite, that Apollo was about. In Scott's mind, the first all-up scientific expedition to the moon would be a testament to one of Scott's favorite quotes, from Plutarch: "The mind is not a vessel to be filled but a fire to be lighted."

Scott and Irwin tear across the Mojave Desert in the Grover during a geology field trip. One geologist said the two astronauts had received the equivalent of a master's degree in his specialty by the time they left earth.

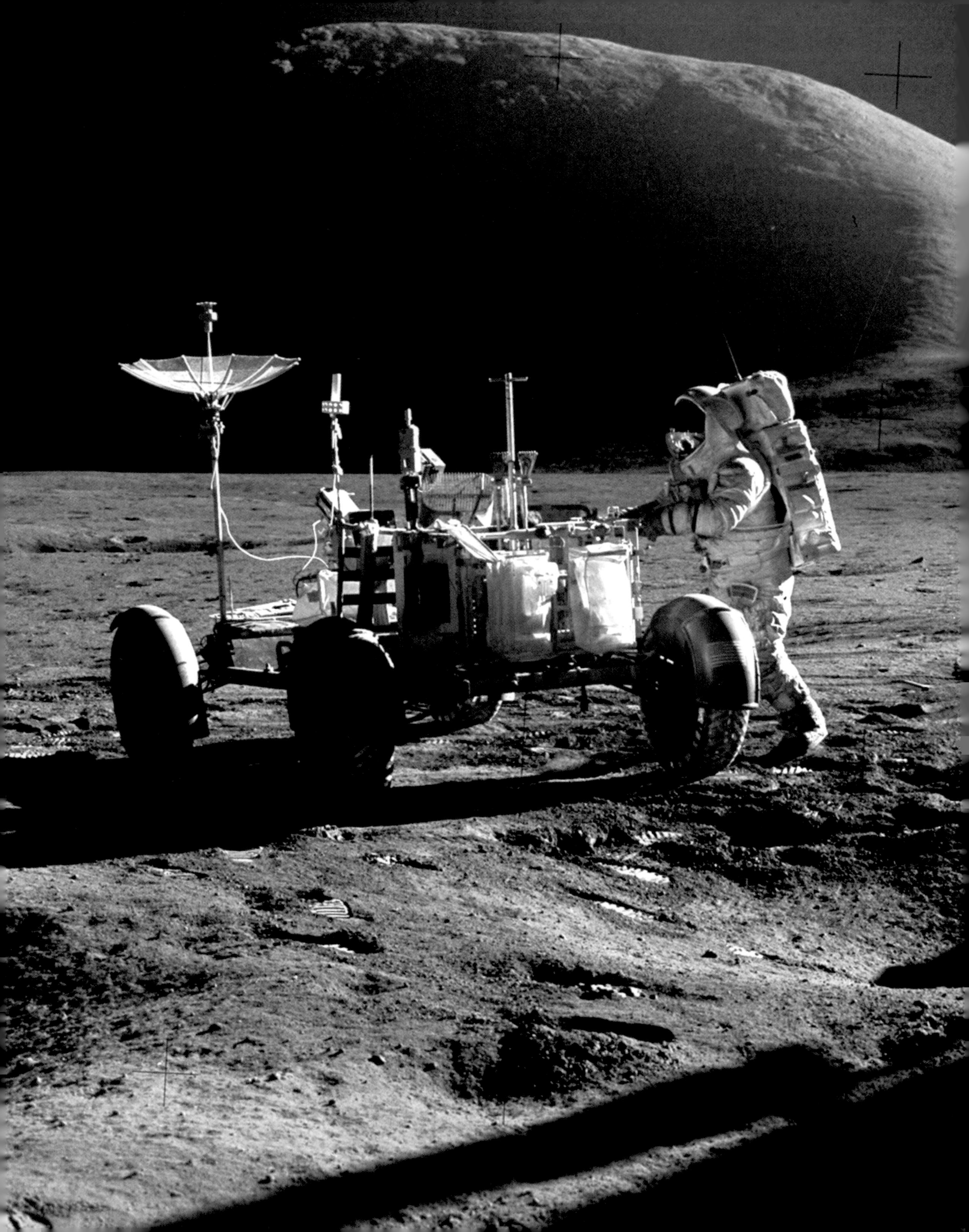

TO THE MOUNTAINS OF THE MOON

APOLLO 15

I: "EXPLORATION AT ITS GREATEST"

Emerging from shadow, the 15,000-foot summit of Mount Hadley looms beyond Jim Irwin as he stows equipment aboard the Rover. Mission commander Dave Scott took this photograph early in Apollo 15's first moonwalk on July 31, 1971.

Of all the dreams that might have visited an eighteenth-century sea captain, could he have envisioned a spaceship sailing a sunlit void in perfect silence, circling the moon? Could he have imagined its commander, a man who, like him, was skilled in pilotage and navigation, but whose energies were fired by the promise of scientific discovery? No doubt if by magic the two men could have stood face to face in the wood-paneled den of some timeless Explorer's Club, they would have had much to say to one another. For his part, Dave Scott had read up on Captain James Cook, the Englishman whose voyages of discovery preceded Apollo 15 by two centuries. In 1768 Cook ventured to the unknown reaches of the South Pacific on the first true scientific expedition, aboard the ship he christened Endeavour. *After the mercantile journeys of Columbus, Drake, and other pathfinders, Cook followed a new lure: knowledge. Though his attempt to witness a rare transit of Venus early in the journey was foiled by haze, Cook went on to map the unexplored lands of Australia and New Zealand. He returned to England in 1771, exactly two hundred years before Apollo 15 left earth.*

If the parallels between these explorations were clear to Dave Scott, he was

equally aware of the differences. In two centuries the scope of exploration had changed dramatically, and with it the role of the explorer. Cook had played a part in everything from commissioning the boat to finding a crew to charting a course. Getting to the moon required the combined efforts of hundreds of thousands of people for the better part of a decade. Cook and his crew were completely on their own from the time they left port until they returned; the crew of Apollo 15 could talk to mission control almost every step of the way. Scott had no illusions about being a modern-day Cook—there simply were none—but as the command module Endeavour *slipped into lunar orbit on July 29, 1971, he could still take inspiration from Cook's words: "I had the ambition to go not only farther than man had gone before, but to go as far as it was possible to go."*

FRIDAY, JULY 30, 1971
5:13 P.M., HOUSTON TIME
4 DAYS, 8 HOURS, 39 MINUTES MISSION ELAPSED TIME

Riding the invisible flame of its descent engine, the lunar module *Falcon*, laden with scientific cargo, cleared the crests of the Apennine mountains and headed for the surface. With less than a minute to go until pitchover, Dave Scott readied himself for the ultimate flying challenge of his career. Just the night before he had told Jim Irwin, "I'm ready. I'm ready to put that baby in there right now." But now, as *Falcon* cleared 9,000 feet, Scott looked to his left and saw something that took him by surprise: the bright flank of Hadley Delta mountain, rising above him into the black sky. He had never seen this view in the simulator. For a brief moment Scott lost the feeling of powered flight and had the unreal sensation of floating slowly past the mountain. Craning to look through the triangular window for a glimpse of the land ahead, Scott saw no sign of Hadley Rille. He was sure they were off target. And moments later, Ed Mitchell in Houston radioed that there was an error in the lander's flight path; they were coming in 3,000 feet to the south.

Suddenly, right on schedule, *Falcon* pitched over, and Scott was confronted by a landscape he did not recognize. Craters he had been memorizing for months —Index, Salyut and a dozen others that were so clear on the Lunar Orbiter photographs—were nowhere to be seen. Instead he saw a nearly featureless, sun-drenched plain. But now, in the distance, he could see Hadley Rille, as distinct as a trowel's cut through wet clay. There wasn't time to go hunting for any other landmarks. Scott used the canyon as his marker and aimed a good distance short of it. He steered to the right, taking out the error in the automatic guidance, and brought *Falcon* in on instinct. Lower now, he could begin to see features he recognized, and at last, a good landing area. In his earphones, he heard

Hadley Rille appears as a dark gash across the face of the moon in this view, made with the lunar module's onboard movie camera as Scott and Irwin began their final descent. The rille served as a guidepost for Scott as he steered the craft toward the landing site.

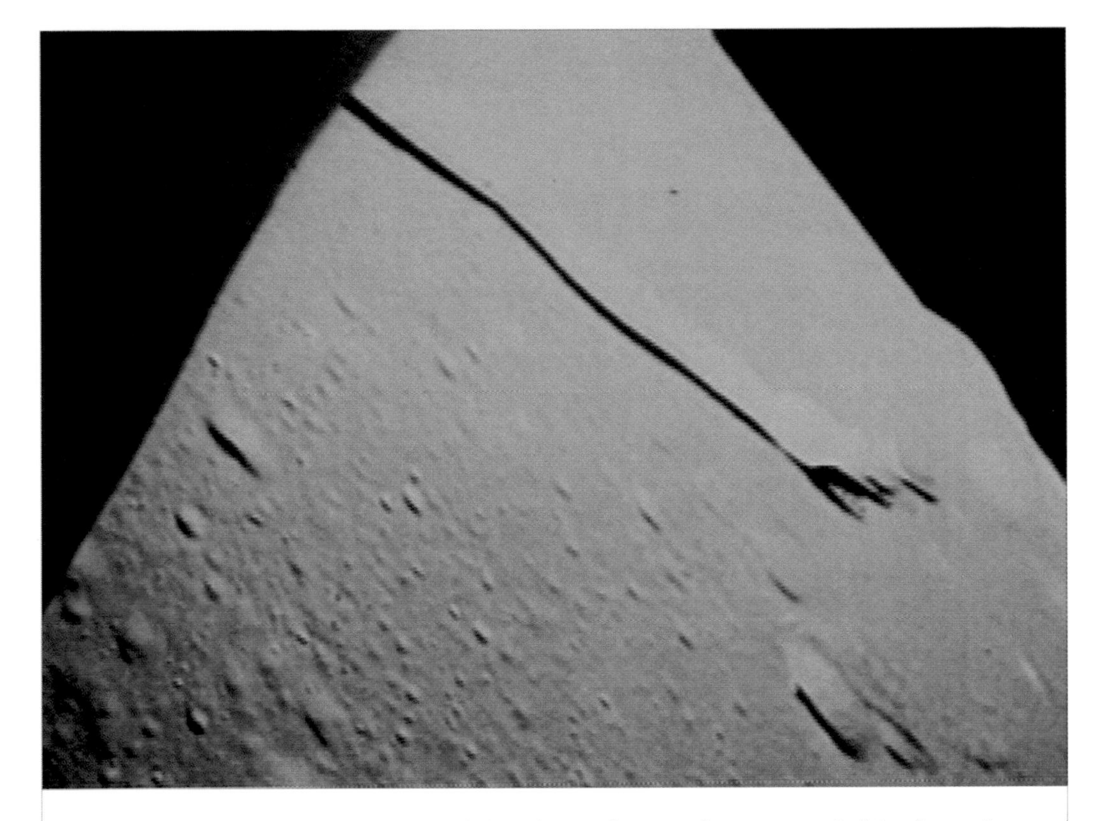

the calm voice of his lunar module pilot with a steady stream of altitude readings and LPD angles: "One thousand feet. Four-five. Nine hundred. Four-five."

Scott gave himself a silent reminder: *Bring it on down.* He'd studied the landings of those who went before him, and he knew that for some reason, perhaps out of an unwillingness to really trust their own landing radar data—and undoubtedly, because it was so hard to judge one's altitude above a moonscape by eye—Conrad and Shepard had unwittingly descended along a stair-step pattern that used up fuel. Scott wanted to avoid that.

At 150 feet he began his final, vertical descent. Immediately he saw streaks of dust blow across the pockmarked ground. At 60 feet a dust storm raged beneath him, completely obscuring the surface. He turned his gaze away from the window and continued on instruments, just as he had done many times in the simulator. He slowed *Falcon* to a near hover, and as Irwin read off the diminishing altitude Scott waited for the Contact Light, mindful of *Falcon*'s bigger exhaust nozzle, which increased the necessity to shut down the engine at the moment of touchdown. When Scott saw the blue glow of the Contact Light, his hand went immediately to the ENGINE STOP button. Several thousand pounds heavier than previous landers, *Falcon* fell to the moon with a firm thud that rattled every piece of gear in the cabin and brought an exclamation of surprise from Irwin: "*Man!*"

The ship pitched backward slightly and tipped to one side. Then there was stillness. "Okay, Houston," Scott announced, "the *Falcon* is on the Plain at

Hadley." As the dust settled the men saw that they had come down in the right place: straight ahead stood Bennet Peak, christened for trajectory expert Floyd Bennet, who helped devise the landing profile that brought them safely over the mountains. Somewhere, much nearer, was Hadley Rille, but it was hidden by the hummocky terrain. From now on, Scott told Irwin, they would call this place Hadley Base. ☾

Scott and Irwin would not go outside now; they would take their first moonwalk tomorrow, after a good night's rest. That was unlike any mission before theirs, but Scott had realized from the first that with 67 hours on the surface, he and Irwin could not do business the way previous crews had. There was an important difference between staying on the moon and *living* there. To perform at their best they would need to maintain their normal day-night cycle, and so, the previous autumn, Scott pushed to change the flight plan. Not only was the new schedule in tune with his and Irwin's circadian rhythms, it opened up some time for something Scott really wanted: a reconnaissance. Lee Silver had taught him: when you arrive at a new field site, go to a high place and look around. Scott was ready to do just that, and the high place was the lunar module's top hatch.

Two of the lunar module's communications antennas frame a quadrant of the lunar horizon in this panoramic view, which continues on the following pages. Scott, whose shadow accounts for part of the dark void below the S-band antenna assembly below, stood in *Falcon*'s top hatch a few hours after touchdown on July 30 to take the pictures from which the panorama was assembled.

As the last wisps of oxygen left *Falcon*'s cabin, Irwin stood by while Scott opened the top hatch, climbed onto the ascent engine cover, and stood up, so that his helmeted head and shoulders stuck up above the LM's gleaming metal structure. From this high vantage, Scott beheld a scene that was at once alien and uncannily familiar. All around him, beyond the undulating *mare*, stood the rounded peaks of the Apennines. Nothing, not months of study, not even the view from orbit, had prepared him for the majesty of the lunar mountains. Their smooth, bright forms were arrayed in fluid sculpture under the black sky. Their slopes were virginal, marred by only an occasional small crater. For eons they had stood unchanged; now, gazing on their ancient beauty, Scott was all but overwhelmed.

Immediately, he spotted familiar features. To the northeast, a slender crescent of sunlight delineated the ridge crest of Mount Hadley, almost 3 miles high, the bulk of the great mountain still cloaked in shadow. Due east, beneath the

sun's obliterating glare, stood the hills he and Irwin had christened the Swann Range. Looking southeast, Scott spotted the lone rocky pinnacle they had named for their professor: Silver Spur. None of these places were accessible to him and Irwin, but to the south lay their destination for the first two moonwalks: Hadley Delta mountain. Over 11,000 feet high, Hadley Delta would have held its own among the Rockies; here it towered imposingly above the *mare,* its steep sides topped by a wide, flat crest. Before the mission, some of the geologists had warned that Hadley Delta's flank might be littered with boulders; no one could predict whether he and Irwin would be able to drive there. Now, thankfully, Scott couldn't see any boulders at all. Tomorrow, he and Irwin would drive onto the western slope to St. George crater, a gouge the size of twenty-seven football fields in the mountainside. Farther along the flank to the east was a much smaller pockmark named Spur, one of the prime targets for the second traverse on Sunday. Monday's plans included Hadley Rille, and while the canyon was still out of view, Scott could see another destination for the final moonwalk, an intriguing collection of craters called the North Complex. Scott would have been amazed to find so many spectacular features in one place on earth, let alone another world.

Scott aimed a telephoto lens at Silver Spur, the North Complex, and a handful of other lures. He rattled off descriptions, speaking not only to Joe Allen in mission control, but to Lee Silver, Gordon Swann, and the rest of the scientists in the geology back room. Then, half an hour after emerging into the sunlight, Scott withdrew into the safety of his lunar module, sure that he and Irwin would not find this place wanting.

10:32 P.M.

Jim Irwin let his remarkably light body settle into his Beta-cloth hammock. He was grateful to be in one-sixth g, glad to be free from the mild but persistent disorientation he'd felt in weightlessness during the trip out to the moon. He could not have wished for a more comfortable bed; his 25-pound form barely sank into the cloth strip. For the first time, no bulky space suit would compromise an astronaut's sleep on the moon. The newly designed suits not only offered more mobility, to aid the work on the surface, but were easier to put on and take off, affording him and Scott the luxury of stripping

down to their long johns and placing the suits, like stowaways, in the back of the tiny cabin. Before the flight, they'd trained to sleep by spending a night in the simulator, with tapes of the LM's coolant pumps and machinery playing in the background. Now, just in case, they wore earplugs. Neither man talked as he lay in the darkened cabin, knowing it was important to fall asleep as quickly as possible.

For just a few minutes, Irwin was alone with his thoughts. He felt lucky to be here; not too long ago he had doubted this day would come. Less than a year ago, he had been so troubled that he had considered resigning from the crew. The grind of training had stressed his marriage almost to the breaking point. His family life had been turned into a series of long separations interrupted by weekends at home that were spoiled by arguments and cold silences. He wanted to focus his entire being on the mission, but his troubles at home distracted him.

This television image from the moon shows Dave Scott as he steps off *Falcon*'s front footpad on the morning of July 31. "Man must explore," said the astronaut as his boots first stirred the lunar dust. "And this is exploration at its greatest."

It was different for Scott; he and Lurton had an air force marriage, and she understood the priorities. She'd been through the training grind with Dave four times before this, counting his stints on backup crews, and by now the family was used to his long absences. But Mary Irwin had never been exposed to military life before she met her husband. As an astronaut's wife she distrusted the fame that went along with the job. She kept her distance from the other wives, forgoing the afternoon get-togethers and the social events of the Astronaut Wives Club. But unlike her husband, Mary Irwin was no loner, and she sought support at a neighborhood church. She and Jim belonged to different Christian sects, and their religious differences were driving yet another wedge between them. By Christmas, with the mission less than eight months away, Mary was on the brink of divorcing him.

More than once, Irwin had considered going to Deke Slayton and asking to be taken off the crew, but each time he'd thought better of the idea. Instead, he took the problem to his commander. He could not have known that his crewmate, Al Worden, had done the same thing when he faced the breakup of his own marriage in 1969. Do what you need to do, Scott had told Worden, promising to talk to Slayton on his behalf. And in fact, Worden stayed on the crew after his divorce. But when Irwin went to Dave Scott, his advice was to wait it out. "Everybody goes through it," Scott said. "It'll change." And it did change; the problems at home began to ease in the last few months before the flight. Relieved, Irwin was free to devote his entire energies to getting ready. Meanwhile, as launch day approached, it was Scott who seemed to grow more irritable, more tight, under the pressure of command.

Last Monday morning, as Irwin lay inside *Endeavour* and felt the Saturn V rise from the earth, he experienced a feeling of release so strong that he almost shed tears: no one could call him back now. Minutes later, when *Endeavour* reached orbit, Irwin was amazed to look out and see the crescent moon, perfectly framed in his window. He was sure it was a good omen.

Now, before he closed his eyes, Irwin did what he always did on earth; he prayed. He gave thanks that he and Scott were here, that everything was working so well. He did not think of his family now, or of events at home; he would say later that he had divorced the earth. He prayed for the success of the mission they were about to begin. Then, Irwin fell into a sound sleep, his best sleep of the whole flight.

SATURDAY, JULY 31
8:29 A.M., HOUSTON TIME
4 DAYS, 23 HOURS, 55 MINUTES MISSION ELAPSED TIME

If Dave Scott's heart pounded as he hopped down *Falcon*'s ladder, it was not really at the thought of becoming the seventh man to walk on the moon; he would say years later that he had already accepted being here. His elation, as a newly minted lunar field geologist, was at arriving at the best field site he could have imagined. As he took his first steps at Hadley Base his words, composed with history in mind, were a preamble for what he and Irwin were about to do: "Man must explore. And this is exploration at its greatest."

The key to that exploration, Lunar Rover 1, was folded up against *Falcon*'s side like a toy in a matchbox. Within minutes, Irwin had joined Scott, and the two men set about bringing the Rover to the surface. They pulled on a pair of lanyards, and the chassis lowered slowly like a drawbridge. Suddenly wheels of wire mesh, with orange fenders, popped out from the corners. With a bit more pulling—interrupted when Irwin slipped and tumbled, laughing, onto the soft powder—the first manned, motorized vehicle was on the moon.

Irwin salutes the flag at Hadley Base. The mountain behind him is Hadley Delta, part of the lunar Apennine range. He and Scott ventured to the base of the 11,000-foot peak during their first and second moonwalks.

9:22 A.M.

It would have been a strange sight, had anyone been there to witness it: a four-wheeled craft heading out among the craters, bearing two space-suited men. The Rover was less a car than a spacecraft on wheels, with its own navigation computer, communications system, and cargo space for such essentials as maps, geology tools—and moon rocks. For all its sophisticated technology, there was still a hobby-shop look to this moon car, with its antennas and folding lawn-chair seats and tool racks. It had been designed, built and tested by Boeing in less than two years; the success of the mission rode on it.

Aboard Rover 1, the ride was more exciting than anyone, including Scott and Irwin, had banked on. Though the Rover averaged only 5 to 7 miles an hour, it seemed more than fast enough to the two men perched atop it. Every time the Rover hit a bump—and the "plain" at Hadley was all bumps and hollows—it took flight. Each new obstacle set one or two wheels off the ground for a long moment. As an added encumbrance, the steering on the Rover's front wheels was inexplicably out of commission, forcing Scott to rely

Scott drives the Rover, his right hand resting on a T-shaped control stick. At the Rover's front end are an umbrella-shaped communications antenna and a television camera, remotely controlled from earth; at the back is a storage rack for geology tools.

on the rear-wheel steering alone. The ride was especially harrowing for Jim Irwin, who felt as though he were strapped onto a bucking bronco, especially when Scott had to swerve to avoid a crater. The transmissions from the Rover were punctuated every so often with a warning from Scott—"Hang on!"—followed by Irwin's grin-and-bear-it laughter. ☾

Scott and Irwin were having as much trouble gauging distance on the airless, treeless moon as everyone had before them. But they were in no danger of losing their way, thanks to the Rover's navigation system; it clicked off readings of heading and distance as Scott drove. Irwin, meanwhile, followed along on the photomap. Aside from the usual difficulty in recognizing landmarks, the maps were misleading: Many of the craters and bumps that looked so prominent on the over-enhanced photos were barely noticeable on the real moon. But one landmark was impossible to miss, and as the Rover headed southwest, over the undulating *mare,* both men suddenly caught sight of it: "There's the rille!" For a time, they followed the canyon's curving edge, then stopped at Elbow crater to gather samples. Soon they were driving again, heading uphill toward St. George; Scott could feel the difference.

11:10 A.M.

"Oh, look back there, Jim—look at that! We're up on a slope, Joe, and we're looking down into the valley, and—That is spectacular!" Standing beside the Rover, Scott and Irwin looked down the mountainside and saw Hadley Rille winding into the distance, its wall curving through shadows and sunlight. They could see boulders the size of houses strewn on the valley floor. Even as he and Irwin set to work, Scott was still reacting to the sight. It was ironic: he'd studied this place so thoroughly he almost felt he knew it in advance; he'd even driven over it in simulations. He'd looked forward to the work he was about to do. But there was one thing about Hadley he had never anticipated, its spectacular beauty. ☾

11:18 A.M.
SCIENCE OPERATIONS ROOM, MANNED SPACECRAFT CENTER

"We have a view of the rille that is absolutely *unearthly.*" Even as Joe Allen's voice sped moonward, it was heard by a roomful of exhilarated geologists in a small command center across the hall from mission control. Here in the Science Operations Room, otherwise known as the geology back room, all

With Hadley Rille as a spectacular backdrop, Scott works at the Rover near St. George crater during the first moonwalk. Tiny dots visible on the floor of the winding canyon are actually house-sized boulders.

eyes were riveted to the television monitors suspended from the ceiling. The color pictures were the sharpest ever transmitted from the moon. Lee Silver felt an excitement that was hard to describe. This was not vicarious pleasure. In essence, he was a participant. He had given some thought, years ago, to the astronaut program, but had decided against it. For one thing, he was almost too old to apply; for another, he knew that, ironically, the total commitment necessary to go to the moon would likely rob him of the opportunity to study the rocks he brought back. Now he and the other geologists were here, not as spectators to thrill vicariously to their students' adventure, but to participate in the exploration itself.

All morning, Gordon Swann and his Surface Geology Team had followed the unfolding moonwalk with the precision of battlefield commanders. One

All morning, Gordon Swann and his Surface Geology Team had followed the unfolding moonwalk with the precision of battlefield commanders.

man, wearing headphones, took notes on index cards, jotting down every rock collected, every picture taken, every comment pertaining to geology. Another listened to the astronauts' descriptions and made annotated sketches, which were then thrown onto the wall by an overhead projector. Another marked down the astronauts' positions on a map that was projected onto another wall. And along the walls were panoramic views of Hadley Base, captured in Polaroid pictures taken from the television monitors. Presiding over this group was another familiar face: Jim Lovell, who had taken charge of the geology back room at Dave Scott's request.

But there was another presence in the back room, and it was a reminder of discord. How to collect rocks on the moon had been a matter of heated debate within the scientific community. To the geochemists, a rock's context wasn't important; all that mattered was getting it back to their laboratories. The geochemists would have been happy to have the astronauts simply fill up a gunny sack with as many rocks as possible. Some even opposed giving them any geology training at all, afraid that the astronauts would go to the moon with the preconceptions of their instructors. On the eve of the mission they had warned the geologists in no uncertain terms that they better not have biased Scott and Irwin. Now, one of them sat in the back room, watching Silver and the rest of the geology team.

Silver had never been able to convince these scientists what he and the others were trying to do with Scott and Irwin. This was not an exercise in programming, but in learning. When Silver showed them anorthosite in the San Gabriels, he made a point of saying that no one knew what the moon's primordial crust was made of. No one knew what they'd find. And the astronauts understood that.

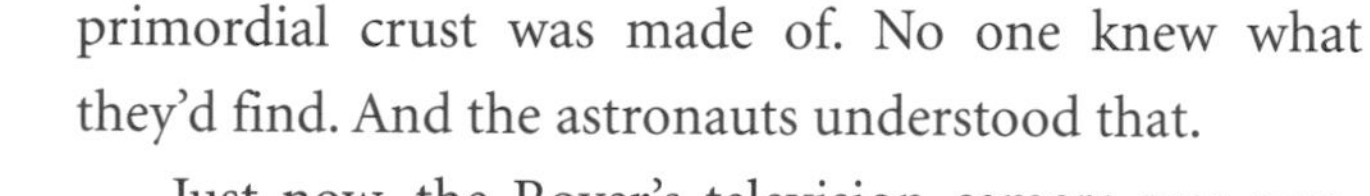

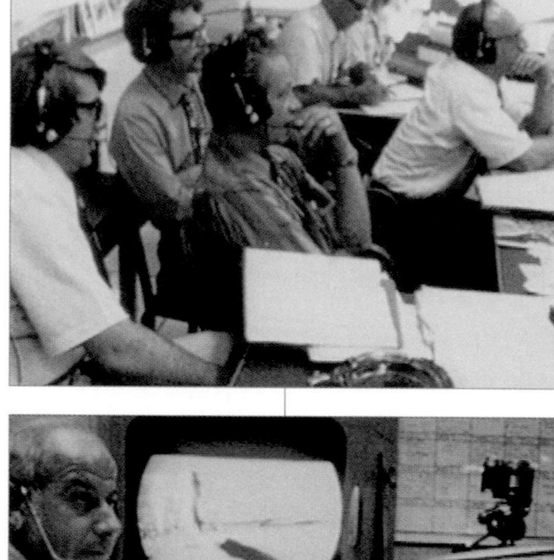

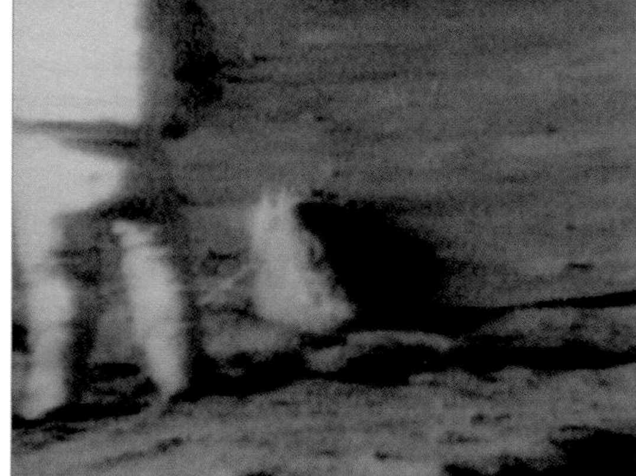

Just now, the Rover's television camera was panning under remote control to show Scott and Irwin at work. There was Dave Scott, standing near a dark, sharp-edged boulder, about knee-high. Scott stood back, describing it before he disturbed it. Silver knew that Scott knew the importance of finding large boulders. Never before had the geologists had the luxury of knowing exactly where a sample had come from. This rock was big enough; it had probably been just where it was for a long time.

Scott and Irwin were in the process of scooping up some of the dust around the rock. There was glass splattered on top of the rock; Scott managed to pull some of it off. Then Scott drew his hammer and pre-

pared to knock a piece off the boulder. He told Allen, "You know what we're going to do when we get through with this thing, Joe? We're going to roll it over and sample the soil beneath." That sample, which had been shielded from the effects of solar and cosmic radiation, would let the scientists determine how long the rock had been sitting there. And once he and Irwin had hammered off pieces of the boulder, which turned out to be composed of two different types of rock, Scott did just that. In this one stop, he and Irwin gathered samples that would meet the requirements of several different groups of lunar scientists. Scott had recognized the optimum collecting site on his own; none of the geochemists, Silver knew, could have prepared him to do that. And when he pushed over that boulder, he felled many of the unspoken doubts that had hung over the back room. ☾

As Scott examines a boulder on Hadley Delta *(below)*, geologists watch from the Science Operations Room *(opposite, top)*. For this front-row seat, the scientists—Gordon Swann in the center of the picture and Lee Silver toward the right—are indebted to engineer Ed Fendell operating the Rover's TV camera *(opposite, center)*, which relays a view of the boulder after Scott has tipped it over *(opposite, bottom)*.

As Scott and Irwin headed back to *Falcon*, the geology team was on a high that would continue long after this first moonwalk was over. History would record that in the evening Dale Jackson of the USGS dined with Joe Allen, Dick Gordon, and several friends at a restaurant called Eric's Crown and Anchor, near the space center. Jackson made no attempt to contain his enthusiasm for Scott and Irwin's performance on the moon.

"Did you see those guys today? They got up there on the side of that mountain and found that boulder"—his booming voice filled the room—"and they sampled the soil around the rock, and then they knocked a piece off it, and then they rolled it over and got some of the soil underneath it!" By

this time people at the next tables had turned around to look. Everyone heard Jackson proclaim, "Why, they did everything but fuck that rock!"

12:33 P.M., HOUSTON TIME
HADLEY BASE

The sweet taste of fruit, a cool swig of water: sustenance for a long day's work. The fruit stick and the water bag inside Scott's suit was a reflection of the fact that no one had ever been at work in a space suit for so long. He and Irwin sorely needed the boost, because now, 4 hours after setting foot on the moon, they returned to *Falcon* to unload their ALSEP package and deploy it on the rolling plains. But before the work could even begin, Allen had some bad news. For reasons that would never be known, Scott's oxygen usage had been higher than expected—in part because of the demands of steering Rover 1 safely among the craters. The moonwalk would probably have to end early.

But for Scott, the hardest work of the day was still ahead. Apollo 15's ALSEP had a new experiment, a pair of thermometers designed to measure the amount of heat flowing from the interior. Scott's task was to bore a pair of 10-foot holes using a battery-powered drill, then insert the sensors. While Irwin set up other instruments, Scott activated the small, boxlike machine

Scott reaches for the battery-powered lunar drill. Boring holes into the moon to place sensors and take core samples consumed much of his time and energy during the moonwalks. At right in the picture is a rack containing sensors to measure the flow of heat from the moon's interior.

that had a handle on either side and a long tube extending from the bottom. Drilling proved to be an unexpected struggle. Shy of the 10 feet requested for the heat flow probes, the drill met resistance and would go no further.

The drill took its toll on Scott. Despite their greater freedom of movement, the new suits had one real weakness, the gloves. No one had been able to develop a pressure glove that was easy to operate. At rest, the glove assumed a position that was half opened, half closed. In order to grip an object like a hammer or a pair of tongs you had to exert constant pressure to keep the glove closed around it. After a few hours of this the forearms ached, but even worse was the damage it did to the fingers. Before the mission, in the interest of having as much dexterity as possible, Scott and Irwin had arranged to have the arms of their space suits shortened so that their fingers were right against the tips of the gloves. When they extended their arms or brought them in close, the rubber tips pressed against their fingernails. At first it was just a minor discomfort; after a while it felt like someone had hit their fingers with a hammer. But for the most part, Scott's excitement eclipsed his pain. And as this first excursion came to an end after some 6 hours, too short in Scott's mind, he stole a last chance to pick up samples. "Oh, look at what I got," he called to Irwin, who was already back in the lander. And then, treasure in hand, he ascended *Falcon*'s ladder, a geologist in a remote and spectacular place returning to the comfort of his field tent.

II: HIGH POINT

SUNDAY, AUGUST 1, 1971
HADLEY BASE

The mountains of the moon captivated Jim Irwin, just as he knew they would. That anticipation had been with him yesterday as he followed Scott down *Falcon*'s ladder. When he put his weight on the footpad, it tipped so suddenly that he went reeling backward, hanging on with one hand, until he was gazing almost straight up at the sparkling blue world he had left behind. Regaining his balance, he took his first buoyant steps into the valley, and could not suppress his elation. The Apennines were not gray or brown as he had expected, but golden in the light of lunar morning. Their rounded, dust-covered forms reminded him of ski slopes—Dollar and Half Dollar mountains at Sun Valley. He was surprised to find that a place others had called stark and desolate could feel so friendly. The mountains surrounded Hadley Base like a cradling hand. He felt at home on the moon.

But yesterday's exploration had taken a toll on Irwin. Not until he and Scott were back in the LM did Irwin feel the full measure of his exhaustion.

Aside from the mental fatigue brought on by hours of intense concentration, and the pressure of finally doing the real thing instead of just simulating it, he was physically spent. The incessant glare from the sun and the bright landscape had given him a fierce headache. And he was parched; the water bag in his suit had refused to work and he'd gone more than seven hours without a drink. Worst of all was the pain in his fingers. By the end of the moonwalk they were so sore that when it came time to take off his gloves he had to ask Scott for help. When they had settled in for the evening, Irwin used a pair of scissors to trim his fingernails all the way back; that helped. He suggested his commander do the same, but Scott declined, reluctant to do anything that might compromise his dexterity.

After the moonwalk Irwin had been overtired, and he had not slept well. Then, this morning he and Scott were awakened early by a call from mission control, to check on a water leak in *Falcon*'s cabin. When they looked behind the ascent engine cover, they found a couple of gallons of water that had leaked through a broken filter. Irwin was glad that Grumman had waterproofed the LM's electronics. Thankfully, despite all of this, Irwin felt renewed as he and Scott prepared for the second traverse. Yesterday, they'd

High on Hadley Delta's flank, Irwin photographed this panoramic view of the Hadley-Apennine valley. Mount Hadley, at center, is more than 10 miles distant. In the continuation of the panorama on the next two pages, Scott photographs a small rock before picking it up and depositing it in a sample bag.

been denied the chance for real prospecting up on Hadley Delta; now they would have one more chance to find out what the mountains of the moon had to offer.

8:07 A.M., HOUSTON TIME
5 DAYS, 23 HOURS,
33 MINUTES MISSION
ELAPSED TIME
SCIENCE OPERATIONS
ROOM

Rover 1 did its drunken gallop across the *mare,* heading south, and in the science back room, Lee Silver listened to Scott and Irwin's progress. As usual, Scott was going full throttle and they were making good time. They zipped past Dune crater, a yawning, block-rimmed hole in the *mare* that was their last landmark before Hadley Delta. Then they reported they were going up a slight incline; they were on the flank of the mountain. With Scott concentrating on his driving, Irwin was doing most of the talking, and Silver was glad for what he heard.

Only a few days before launch, after the last of a regular series of evening geology briefings at the Cape, Silver had said good-bye to his students. In the parking lot of the crew quarters, he wished them well. He felt admiration for

them, and also concern, but they radiated confidence. He turned to Irwin. Though Irwin was just as adept as Scott, Silver knew, he usually deferred to his commander and said comparatively little. One day late in the training he'd taken Irwin aside and said, "Talk to me up there, Jim. Don't just follow Dave around." Now, saying his good-bye, Silver offered one more smiling admonition: "Now, Jim, we're going to hear from you up there, right?" ☾

Irwin smiled back. "You bet."

Silver needn't have worried. Irwin was full of detailed, precise descriptions as the Rover bounced along. He brought a patient thoroughness to the explorations that complemented Scott's drive and exuberance. He frequently directed his commander's attention to something too important to ignore, but always as a suggestion: "Dave, shouldn't we get that rock over there?" And now, as Silver listened, Irwin made a new discovery: The side of Mount Hadley was covered with a slanting pattern of straight lines, as if a giant comb had been dragged across it. Everyone in the back room wondered excitedly whether the lines might help solve the riddle of how these mountains had formed; perhaps Mount Hadley was a block of layered bedrock that had been raised up by the violence of the Imbrium impact. For more clues, they would

have to wait for the photographs. And when Irwin brought Scott's attention to the new find—"Oh, *yeah,*" Scott said—Silver found himself smiling.

8:28 A.M.

The Rover climbed effortlessly; even now, heading directly up the slope, it was making 6 miles per hour. The plan was to work their way along the mountainside, sampling whatever geologic variety Hadley Delta had to offer. Scott knew what the geologists wanted—a crater that had acted as drill hole into the mountain, strewn with boulders torn from the flank. But there were no such craters in sight. There was a sameness to the terrain up here; everything was much more worn and subtle than the photographs had led them to believe. Angling along the contour of the mountain, Scott searched for a target, finally spotting a medium-sized crater. There weren't any boulders, but it would have to do.

When he hopped off the Rover, he almost fell over backward. The drive had been so easy that he had no idea the steepness of the slope they were on. Glancing at his feet, he saw to his surprise that his boots were half-buried in dust, and yet the wire-mesh wheels of the Rover, fully loaded with both of them in space suits, had penetrated only a fraction of an inch. He warned Irwin to be careful. Then he turned around, and what he saw almost knocked him off balance again. They were more than 300 feet up. Hadley Delta's broad flank swept downward and merged with a bright, undulating desert adorned with the brilliant white rims of craters. Beyond, the Apennines invaded the dome of space. The entire valley was one enormous, incredibly clear panorama: It was the face of pure Nature. And out in the middle of this pristine wilderness, a single artifact: his lunar module. *Falcon,* that big metal-and-foil bird, was a mere speck.

The lunar module *Falcon* stands three miles away in this picture, which Scott took with a telephoto lens. Behind the LM is Pluton crater, 2,300 feet across.

No scene could have conveyed more vividly the reach of this exploration—or the risk. If something had happened to the Rover now, Scott and Irwin would have faced a long and difficult walk to safety, for the LM was more than 3 miles away—much more than the distance covered by Shepard and Mitchell during their round trip to Cone crater. The whole question of how far to let two Rover-riding astronauts go had consumed hours of premission deliberation. At no time could the men be allowed to drive farther than they could walk back with the amount of oxygen remaining in their backpacks. Because their oxygen supply dwindled as the moonwalk progressed, this *walkback limit* would be an ever-tightening circle.

Even if the Rover worked flawlessly—and so far, it had done nearly that—there was always the chance that a backpack would fail. In that case, the men would break out a set of hoses that would allow them to share cooling water. The man with the failed backpack would survive on his own emergency pack, which contained about an hour's worth of oxygen, and if necessary, his partner's emergency pack, allowing more than enough time for the

Rover to race back to the lander. A more dire scenario was that both the Rover *and* a backpack might break down. For a time, this remote possibility had so worried the managers that they considered writing the mission rules around it—a change that would have severely limited Scott and Irwin's explorations. In the end, NASA bought the risk of the double failure, knowing that if it came to pass, one of the astronauts would not make it back to the lander alive. But all of this was far from Scott's mind as his gaze lingered on the view. He felt a wave of excitement: everything was working. He joined Irwin and the two men headed down the slope to the crater.

The mountain was a difficult workplace. The stiffness of their suits hindered climbing; they were reduced to taking small, ineffective hops. With every step, the soft, thick dust fell away from their feet, as if they were walking on the side of a sand dune. A few steps left them nearly out of breath. They spent the better part of an hour here, winning only a few samples for their efforts. There just wasn't the variety they'd hoped for. They would make another try at Spur crater, but first Scott wanted to get to a boulder they'd spotted along the way.

"Man, I'd sure hate to have to climb up here!" said an amazed Scott as he struggled back up the slope, grateful that the Rover had done so well. "You'd *never* get here without this thing." And yet, they had barely begun to scale Hadley Delta. With no haze to block the view, its upper reaches were clearly visible, many thousands of feet beyond. Gazing on that bright frontier, Irwin silently wished they could go higher.

9:38 A.M.
MISSION OPERATIONS CONTROL ROOM

Joe Allen sat at the Capcom's console, studying a photomap of Hadley Delta. Every so often he made a mark showing the best estimate of where the Rover was. His brown hair, short on the sides but longer in front, fell across an unlined forehead. Allen listened intently to the voices of his friends on the moon; at the same time he kept an ear tuned to the conversations in the geology back room. He also had a TV monitor on which the scientists could write him notes—"We'd like a core-tube at this stop"—so that if a question came down from the moon, Allen would be ready with the answer. And it was Allen's job to take the myriad inputs to the astronauts, from the geologists, the flight director, and the flight controllers, and synthesize them into one coherent voice.

Joseph Percival Allen IV, an Iowa-born physicist who came to NASA with the XS-11, was sometimes known as Little Joe, because at five feet six inches he had usurped Pete Conrad by half an inch as the shortest man in the Astronaut Office. At thirty-four, he looked as young as the day he entered Yale to begin working toward his doctorate. In the minds of the geologists, Allen's assignment as mission scientist was one of the best things that happened to Apollo 15. Allen combined a youthful enthusiasm with a keen scientific mind and an innate sense of people. On the field trips, it was Allen who took the edge off Scott's driving intensity with a humorous remark. Now, as Capcom, Joe Allen was Scott and Irwin's link with the back room. No one but he could speak directly to Scott and Irwin; any requests, advice, or questions from the back room had to be spoken by Lovell over the loop to flight director Gerry Griffin and then, via Allen, to the moon. It was no accident that every link in this human chain had been out on the field trips at one time or another, and that each understood the scientific objectives of the mission. Allen in particular understood so well that often he did not even need to ask the back room what to say. He was more than just an ally, Lee Silver would say years later; he was a colleague. ☾

Throughout the moonwalks, Allen had tried to convey a sense of support and optimism. He knew Scott and Irwin were working very hard, and he never missed an opportunity to ease the pressure with a comic remark, a private joke. At each new achievement, however minor, Allen voiced his approval: "*Extraordinary. Superb.*" Deke Slayton had come down on him once or twice for deviating from the spare style of communication that characterized standard pilots' radio discipline. Allen didn't let that bother him. There wasn't any operational reason not to let Scott and Irwin know that he was pulling for them.

He knew roughly where Scott and Irwin were: a few hundred meters east of Spur crater, and slightly upslope from it. Scott had already radioed that he didn't think it was worth going any farther to the east; they'd already seen as much variety as they were likely to see. They had stopped to sample a lone boulder that was apparently too good to pass up. Allen listened intently as Scott maneuvered the Rover, then stopped. Suddenly the radio link was scratchy, but through the static Allen could hear Scott's labored breathing. He heard Irwin say, "Gonna be a bear to get back up here, you know."

"Hey, troops," Allen said with contained urgency, "I'm not sure you should go downslope very far, if at all, from the Rover."

"No, it's not that far," came Scott's answer. Irwin added gamely, "I think we can sidestep back up."

There was no television from the Rover just now, but Allen could picture

IN 1971

Patton* wins the Oscar for best motion picture.

The U.S. and the U.S.S.R. begin the fourth round of the Strategic Arms Limitation Treaty (SALT) talks in Vienna, Austria.

Louis Daniel "Satchmo" Armstrong dies at age 71.

William Peter Blatty publishes *The Exorcist*.

Inmates at New York's Attica State Correctional Facility riot and take 40 hostages. National Guardsmen storm the prison, killing 29 convicts.

the steep, powdery slope. He also knew Scott and Irwin wouldn't give up a prize without a fight. And he could sense that the managers in the back row of mission control were getting worried.

Suddenly Scott called out that the Rover was beginning to slide down the hill. "The back wheel's off the ground," said Irwin. Their voices didn't convey the seriousness of the situation: If the Rover got away from them, it meant scrapping much of the mission, and a long walk back to the LM. Allen listened as Scott held on to the Rover to keep it from sliding down the hill. Now he heard Irwin talking about something he'd noticed on top of the boulder, some kind of green material, he was saying. Allen knew that was unusual, and he wasn't surprised when Irwin urged Scott to get a sample. The men traded places; now Irwin held the Rover while Scott made his way to the boulder.

"Use your best judgment here," Allen cautioned once more, sensing now that Scott and Irwin had the situation under control. Within minutes, having bagged a piece of their strange find, the men were driving again, heading west toward Spur crater. Time was running out; this would be the last stop on the mountain.

10:00 A.M.

Heading downslope once more, Scott and Irwin angled toward Spur crater and stopped next to the rim of the football-field-sized pit. Spur looked promising. There weren't many big rocks, but there was great variety in the small fragments that littered the soil. Immediately Irwin noticed a white rock, perched atop a small, light-gray pinnacle, that was different from anything he'd seen. Irwin wanted to rush over to it, but Scott was already collecting another rock. Even as Irwin went to help him, he spotted something even more curious. It was more of that green material, like the coating on the boulder at the last stop. What could it be? Moon rocks were gray, or they were tan, or black, or white—but not sparkling light green. To Irwin, a man of Irish descent who was born on St. Patrick's Day and had shamrocks tucked away in the lunar module, a green rock was a special find. There was some debate about whether the color was real—Scott thought their gold-plated visors were playing tricks on them—but they picked it up anyway. It looked like a chunk of basalt, shot through with tiny holes where gas bubbles once frothed in molten lava. But it couldn't be basalt; it actually yielded to the pressure of Scott's gloved fingers, which left streaks in its surface. Not until the rock was unpacked in the Lunar Receiving Laboratory would the men see

Irwin struggles to keep the Rover, parked on a steep incline, from slipping down the slope while Scott photographs an unusual boulder. Rear wheels in the air, the Rover was in a precarious situation and, in earth gravity, might well have gotten away from Irwin.

that it truly was green, and that it was actually made of tiny spheres of glass. Even among moon rocks, it was a rarity. And in time, the story it would tell, of eruptions from the hidden depths of the lunar interior, would hold geologists spellbound.

But just now, Scott and Irwin's attention was riveted to the strange, white rock Irwin had seen earlier. As they approached, they strategized about the best way to collect it. Irwin suggested lifting it off its dusty pedestal with the tongs. Scott did so, and he raised it to his faceplate to inspect the find, which was about the size of his fist. It looked fairly beaten up, and it was covered with dust, but as he grasped it with his gloved fingers some of the ancient coating wiped off, and he could see crystals: large, white crystals. Suddenly—

"Ahhh!"

"Oh, man!"

—both men realized what they had discovered. The rock was almost entirely plagioclase. Exposed to light for the first time in untold eons, its white crystals glinted in the sun, like the ones they had seen in the San Gabriels. This was surely a chunk of anorthosite, a piece of the primordial crust. Scott radioed the news.

Physicist-astronaut Joe Allen, who served as Apollo 15's mission scientist, manned the Capcom console during the moonwalks. In his transmissions to Scott and Irwin, Allen relayed support, concern, and enthusiasm as needed.

"Guess what we just found. *Guess* what we just found. I think we found what we came for."

"Crystalline rock, huh?" prompted Irwin.

"Yes, sir," said Scott.

"*Yes, sir,*" repeated Joe Allen, like a brother at a revival meeting. After describing the rock in detail, Scott placed it in a special sample bag by itself. Of all the rocks brought back from the moon this one would be the most famous. To the geologists, it would be sample number 15415, but to the world it would be known by the name bestowed upon it by a reporter covering the mission at the space center: the Genesis Rock. In time, probing the treasure with electron beams in their laboratories, the geologists would peg the rock's age at 4.5 billion years. If the moon was any older than that, it wasn't much older; the solar system itself was thought to have formed only 100 million years earlier.

Scott couldn't hear the geologists' ecstatic reactions in the back room, but he didn't need to. And Spur had even more to offer. As they made their way along the crater rim they happened on one remarkable find after another, poking out of the dust. Scott sounded like a rare-coin collector who had stumbled across a treasure chest in the attic. "Oh, look at *this* one!" Scott told Allen what was already clear: "Joe, this crater is a gold mine!"

"And there might be diamonds in the next one," said Allen, a gentle reminder that Scott and Irwin could not linger here. They were getting close enough to their immutable walkback limit that Allen was telling them to press on. Now Allen radioed that Irwin's sample bag was about to come loose—everyone in mission control could see that on the big screen, and they worried that all those priceless samples might tumble into Spur crater. Precious minutes were lost as Scott cinched it up. With less than 15 minutes remaining, Scott and Irwin looked hopefully at a boulder that lay perhaps a dozen yards farther along Spur's rim, the only large rock in sight, and therefore the only rock for which the geologists could be sure of its place of origin. Undoubtedly it had been torn from the floor of Spur.

Just now, Allen passed up word from the back room: forget the boulder; they wanted them back at the Rover, using the rake to collect a bunch of walnut-sized fragments. Scott eyed the boulder. He told Irwin to go ahead and start raking; in the meantime he went into high gear. A short run along Spur's rim brought him to his quarry. Working on stolen time, he clicked off Hasselblad pictures of it from every angle. There wasn't time to hammer off a piece, but he spotted a small fragment in the dust—obviously knocked loose from the big rock—and got his sample after all. He hoped the geologists

Scott advances toward a small white rock perched on the rim of Spur crater. A chunk of the moon's primordial crust, the specimen became known as the Genesis Rock. The rock was more than 4.5 billion years old when the astronauts carried it to earth.

would be able to figure out where it came from by studying the pictures. Scott turned and ran back to join Irwin in his work. He hated to leave. It was a frustration he would feel again and again on the moon, just as he had that first day in the Orocopias with Silver: there was never enough time.

12:41 P.M.

Heading back to the LM, Irwin was glad to be traveling on more level ground. Up on the mountain, there had been times, driving along the contour of the steep slope, when Irwin was afraid the Rover would tip over. Now, pulled away from their explorations by the pressures of time and timeline, he and Scott made a brief stop at the LM, then revisited their ALSEP package, that tiny city of instruments sprawled in the dust. Scott had to summon more strength from his sore, tired hands to drill into the soil for the socalled deep core sample. While Scott drilled, Irwin dug a trench in the gray dust, took pictures of it, and used a device called a penetrometer to test the bearing strength of the trench walls and bottom. Irwin's tests would reveal much about the mechanical properties of lunar soil.

Scott's fingers were already so badly injured that they were beginning to turn black, and they were sore from cinching up sample bags all morning. Using the drill just made them worse. He had to bring his hands in close to his chest in order to squeeze the drill's trigger, and he could stand only a minute or so of the pressure against his nails before he had to back off, shaking his arms against the pain. But he kept at it. Now one core tube was in, then he added a second. He could feel the drill meet resistance, then penetrate more easily again. Soon he had drilled the full 10 feet down. It was the deepest lunar core sample ever; inside were millions of years' worth of lunar history. Now he tried to pull the core out of the ground, but it wouldn't come. When Joe Allen told him to leave it until tomorrow, he was disappointed. Still, it was getting to be a long day, and he could feel it. By the time he and Irwin climbed back inside *Falcon,* they had been outside for more than seven hours. Today alone, they had driven almost 8 miles. And they had been rewarded for their hard work; it was a day in the field any geologist would have been proud of.

The secret of living on the moon, in Dave Scott's mind, came down to one thing: getting out of his suit. He and Irwin would have been miserable if they'd been sentenced to three full days inside them. By now, the whole

PROBING THE MOON FROM ORBIT

The command ship *Endeavour (right)* carried an array of sensors and cameras called the Scientific Instrument Module (SIM), the first of its kind. Installed behind a protective cover jettisoned before entering lunar orbit *(below, left)*, the SIM included a star camera for determining the spacecraft's orientation (1); a laser altimeter for measuring lunar topography (2); and a mapping camera (3) and a panoramic camera (4) for photographing the moon from orbit. In addition there were instruments for studying the all-but-nonexistent lunar atmosphere (5) and for analyzing the chemistry of the lunar surface (6, 7). A small satellite ejected into lunar orbit *(below, upper right)* mapped the moon's gravity field and counted charged particles emanating from the sun.

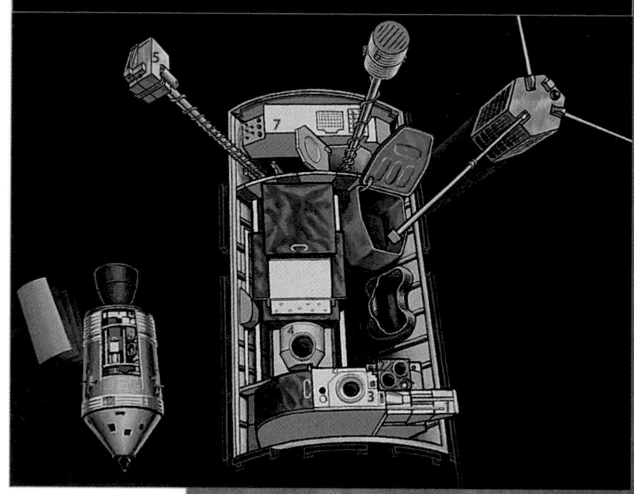

process was routine: Off came the gloves, then the helmets, and then, one at a time, each man stepped into a white stowage bag and pulled it up over his legs to catch the rain of soot, and stripped to his long johns. Piled in the back of the cabin, the suits would dry out from almost nine hours' worth of wear. Then came the chores: recharging the backpacks with oxygen and water, stowing the rocks they had collected that day, and attending to *Falcon*'s systems. And there was another routine that wasn't in the flight plan. Each day, Scott and Irwin paused in their work for a brief chat with their orbiting companion. The signal was broken up until Worden cleared the mountains, but then there was time to exchange news and pleasantries as *Endeavour* sped, starlike, overhead. Scott was glad to hear that Worden and his arsenal of scientific instruments were going like gangbusters. And then, before *Endeavour* drifted out of range, more mundane matters.

"Hey, Al," called Irwin, "throw my soap down, will you? And my spoon."

"You forget something, Jim?"

Irwin grinned through lunar grime. "I really need my soap."

●●◐○○○◑●

Aboard *Endeavour,* everything was still quite clean and neat—except, perhaps, Al Worden, who had at this point gone a week without a shower or a shave and was happy to have Irwin's soap aboard. Like his friends at Hadley Base, Al Worden was too busy to care how grungy he was. He was, in effect, the commander of the first fully equipped orbiting lunar science platform. *Endeavour*'s service module was crammed to the hilt with high-powered cameras and sensors, and much of Worden's time was spent turning the instruments on and off as dictated by the flight plan. Two different sensors were designed to map the chemical composition of the surface. Another would try to detect evidence for volcanic gases seeping out of the moon during recent geologic time. A mass spectrometer was included to sniff out extremely tenuous gases that might surround the moon. There was even a tiny satellite, which Worden would release into lunar orbit in a couple of days, equipped with a magnetometer to stalk the elusive lunar magnetic field. And the two cameras were superb creations. One of them, based on declassified spy-satellite technology, could photograph details on the surface as small as a yard across. It was just possible, Worden thought, that this enormous haul of data would eclipse the knowledge gained from the rocks his crewmates were collecting down there amid the dust. ☾

And to this list of sensing apparatus, add Worden himself, for like Scott and Irwin, he had given himself to the study of the moon. Worden, like Stu

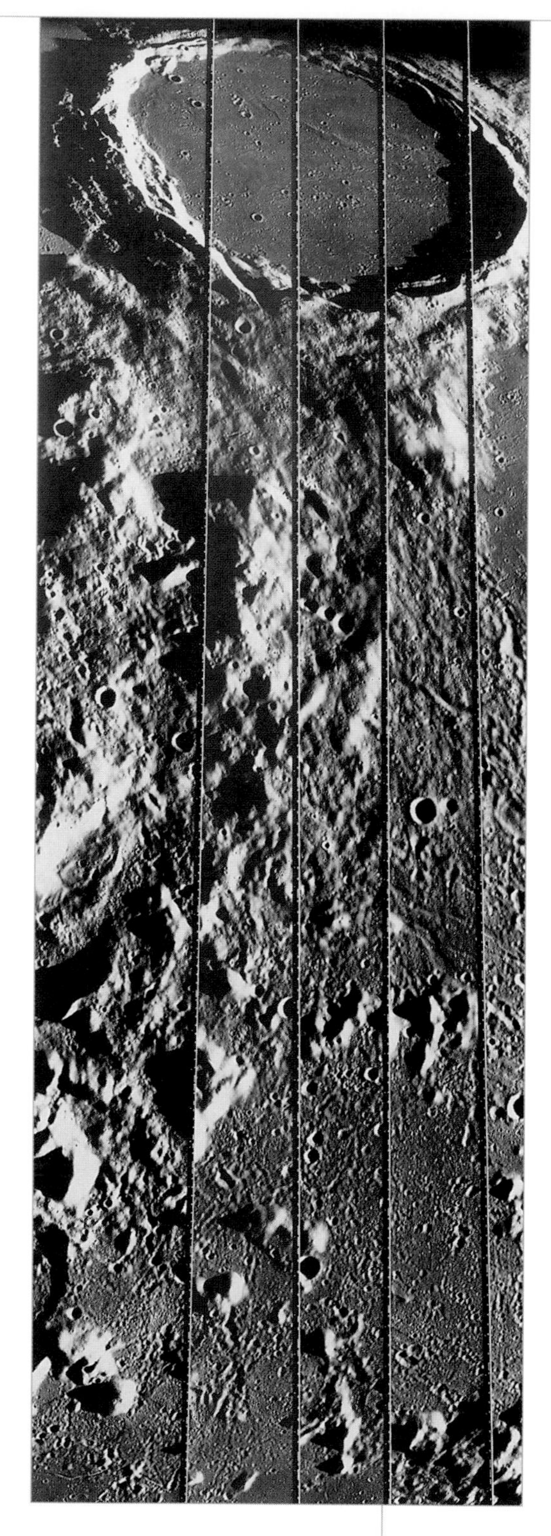

Ribbonlike frames from *Endeavour*'s panoramic camera show the 50-mile-wide Archimedes crater. Mountains of ejecta below it splashed out of the crater when the meteorite that created it struck the moon.

Roosa, had made Farouk El-Baz his mentor. For the first time, a program of visual observations from lunar orbit were an official mission objective. Worden not only admired the exuberant Egyptian—or the King, as the astronauts called him—he liked him as much as he did anyone in the space program, and found him an inspiring teacher. After the study sessions, they would go drinking together. "Now," Worden would say, "we're going to see what a White Russian can do to an Egyptian." It would always turn into a competition, and El-Baz would hang in there with a tenacity any fighter pilot would have applauded.

It was for the King that Worden spent dozens of hours flying over the mountains and deserts of the western United States, honing his observational skills, and many more hours studying the Lunar Orbiter photos. And now, in lunar orbit, each time *Endeavour* flew into sunlight, Worden was amazed to look down and see among the pockmarked chaos a crater he recognized. How strange, to be *here* and see something familiar.

No feature was more memorable than giant Tsiolkovsky crater, and for this and other targets Worden carried photographs with instructions for specific things El-Baz wanted him to watch out for. Looking down on Tsiolkovsky, he could see that it was packed with spectacular geology. The crater was so big that when he was directly overhead it nearly filled his view. The central peak was enormous, a great white mountain rising out of that floor of dark lava. Worden could see layers in the peak, as if it were a great slab that had somehow been turned on its end. Around it, he spotted boulders that must have been the size of a city block. Along the giant's rim, he could see what appeared to be rock avalanches that had cascaded away from the crater onto the surrounding hummocks.

Now he was over a place called Littrow, near the Sea of Serenity. Farouk had told him to look very carefully at Littrow, because in the unmanned photographs, there were craters, surrounded by halos of dark material, that resembled small volcanoes, or cinder cones, as Farouk called them. If that was true, they might represent an entirely different, and possibly much more recent, type of volcanic activity from the *maria*.

Sure enough, Worden could see the small, dark-haloed craters, and he told Capcom Karl Henize excitedly, "It looks like a whole field of small cinder cones down there." On the next orbit he would get another look, and he would report the craters were cone-shaped, and that he was even more convinced they were volcanic, and Henize would say, "Keep talking like that and we might end up going to Littrow sometime."

Like anyone who had ever been here before, Worden's mind was split between the operational tasks that filled his consciousness and the sights, sounds, and feelings that were even now being stored in his subconscious. Only during the orbital night did he have time to savor the experience. He had never believed that the solar system was the center of the universe in any sense; in the back of his mind, he'd always thought that the earth was just one planet among millions. But it wasn't until he found himself over the far side of the moon and saw all those stars—God, there were so many stars—that he was sure there was more to the universe than he had ever imagined. He knew that astronomers had debated the probabilities of intelligent life in the universe, but this light show was enough to convince Worden that it had to be all or nothing: Either there was no life out there, or the cosmos must be teeming with it. ☾

Just before emerging from darkness, Worden looked out ahead and saw waves: thin, sinuous bands of light, suspended between the starry heavens and the unyielding darkness that was the moon. At first, Worden did not know what they were. Then he realized he was seeing the rays of the sun, which had not yet risen, glancing off the tops of distant mountains along the horizon. They were the opposite of shadows. The waves thickened until, without warning, night turned almost instantly to day. As his eyes adapted to the light, Worden searched once more among the craters for a familiar place, and for the places he would try to make a little less unknown.

●◐◐○○○◑◑

Jim Irwin lay in his hammock, awaiting sleep. Before the mission, inspired by the prospect of visiting the lunar mountains, Irwin had felt a desire to hold a short religious observance on the moon, but when he mentioned the idea to his commander, Scott quickly turned him off. But ever since they had arrived on the moon, Irwin felt a spiritual quality about the place. Seemingly insoluble problems had come up, and each time they had been resolved. When he was deploying the ALSEP, he had a problem that had never happened in training; the cord for deploying the central station broke. Irwin prayed for guidance, and immediately he knew the solution was to get down on his hands and knees and pull the cord manually. Then there was the Rover's rear steering, which had been out of commission yesterday, and nothing they did could fix it. Inexplicably, when he and Scott came out this morning, it was working. He felt a glow inside him that whatever problem came up, they would solve it.

Once again Irwin took a moment to pray. He gave thanks that everything was going so well. He gave thanks too for the discovery of the white rock, so remarkably displayed on that pedestal of dust, as if it were being presented to them. Tomorrow would surely hold more discoveries, and before he closed his eyes, Irwin prayed that it might also bring him an opportunity to make his own small spiritual gesture. On the moon, a favorite biblical passage, a verse from Psalms had drifted through his thoughts like a refrain:

I will lift up mine eyes unto the hills:
from whence cometh my help?
My help cometh from the Lord.

He hoped that sometime during the next day's activities he would find the right moment to recite it. As he prayed he sensed that God was near him, even here.

III: THE SPIRIT OF GALILEO

Joe Allen liked to think he was cut from the same cloth as the pilot-astronauts, but working for Dave Scott had been a lesson in humility. The man's stamina was simply amazing. It was beyond him how Scott could go through the long days of meetings and simulations without the slightest sign of wearing down, seemingly as productive and alert at ten o'clock at night as he had been at breakfast that morning. No question, it did wonders for his motivation to see his boss going at it so relentlessly, but sometimes it was all Allen could do to keep up. He was reminded of the climactic scene from one of his favorite movies, *The Hustler*. Jackie Gleason as Minnesota Fats is twenty-four hours into his pool-room confrontation with Paul Newman as Fast Eddie. Just as it seems that Eddie is about to clean the old master out of his last dollar, Fats takes time out. Going into the seedy little bathroom, he washes his face, cleans his fingernails, puts on a fresh shirt and a tie, and emerges into the smoke feeling like a new man. "Fast Eddie," he says, "let's shoot some pool." Fats goes on to win back all his money and win the match. Allen told Scott about it one day, and after that, when Scott would see him flagging midway through a late-evening planning session, he would lean over and say, "C'mon

Joe, let's shoot a little pool." Now, as Scott and Irwin finished suiting up for a final day of exploration, Allen asked Scott if he and Irwin might be ready to shoot a little pool themselves. The answer from the moon: "Joe, today's the day for a little pool." Now came the endurance lap.

MONDAY, AUGUST 2
4:07 A.M., HOUSTON TIME
6 DAYS, 19 HOURS, 32 MINUTES MISSION ELAPSED TIME

Outside, the first men to spend three days on the moon saw the sun almost directly overhead, sending its fierce heat down on Hadley Base. Even within their suits, Scott and Irwin felt its warmth. They had been nearly two hours behind schedule getting to sleep last night and waking up this morning, and that had cost them. The liftoff time, slated for later this day, could not be changed. Allen had already told them that this third moonwalk would be shortened to only four or five hours. Scott and Irwin planned to visit Hadley

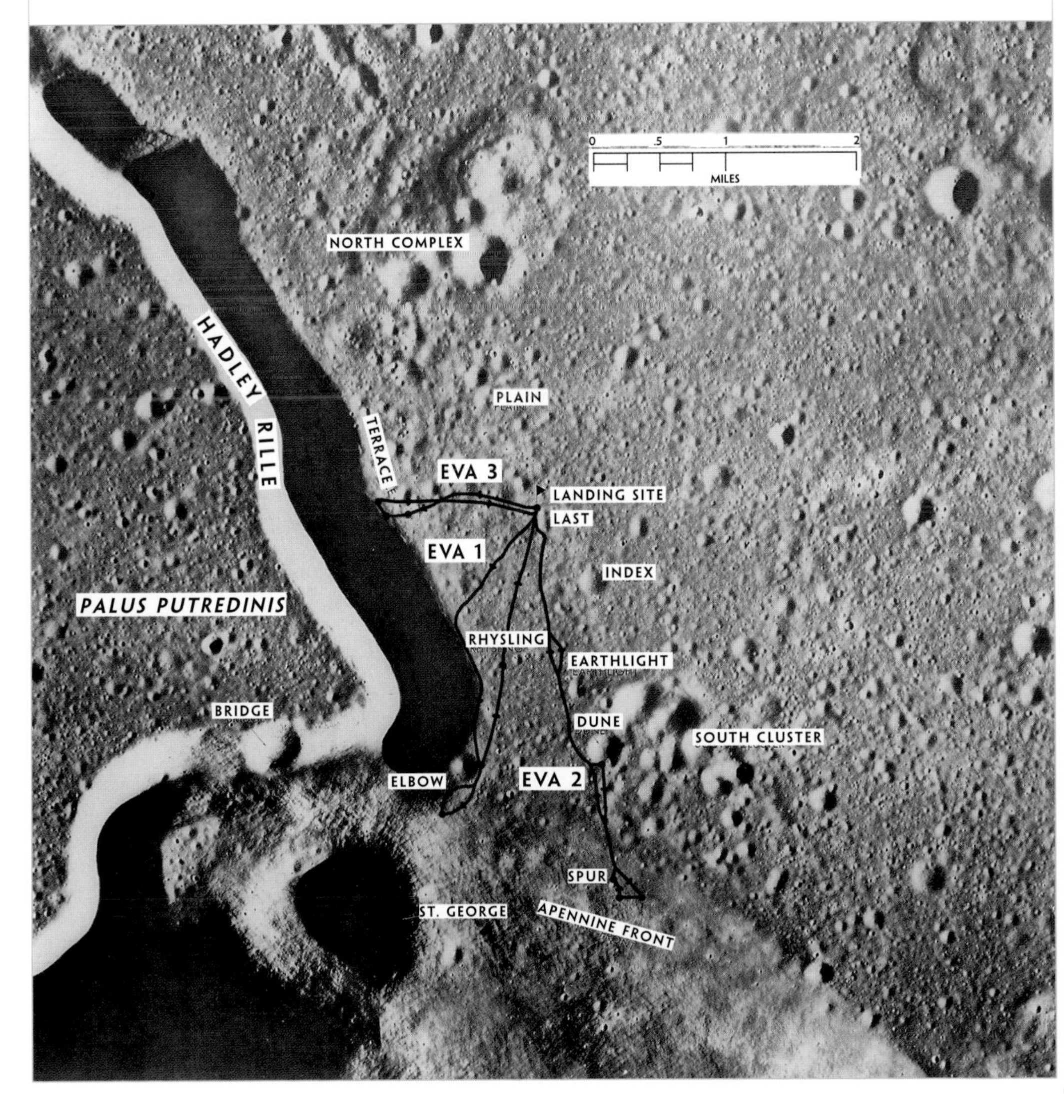

During their three moonwalks—called extra-vehicular activities (EVAs)—on the *Palus Putredinis* (Marsh of Decay), Scott and Irwin followed the routes drawn on the photograph at right. The astronauts had planned to visit the North Complex, a group of suspected volcanic craters near the top of the picture, but they abandoned that goal when they fell behind schedule.

Rille, and then, if there was time, the mysterious group of craters called the North Complex. A young Survey geologist named Jerry Schaber had advanced the idea that the North Complex might actually be a cluster of small volcanoes. Even if they were impact craters, Schaber said, they would still be worth the trip; the largest was almost 2,500 feet across, big enough to have punched through the frosting of *mare* basalt and bring up chunks of the more ancient rocks underneath. The North Complex hadn't been on the original traverse plan, but when Scott heard Schaber's idea, he personally made time for it. Solving that mystery was one of Scott and Irwin's fondest hopes for the mission. But they would have to wait; Allen was sending them to the ALSEP site, for a showdown with Scott's nemesis, the drill.

4:41 A.M.

It was waiting for them when they reached the ALSEP, sticking out of the ground just as Scott had left it the day before, when he had tried unsuccessfully to extract the deep core sample. Scott had already begun to wonder whether the core was worth the time and effort it was costing. Fighting their pressurized suits, the men bent over to grab the drill's handles and pulled as hard as they could. The drill did not budge. Scott gave a Herculean pull—"One, two, *threee*"—and it just barely moved.

In mission control Joe Allen listened with amazement. Scott and Irwin represented as much physical strength as any two men in the astronaut corps. In training, the drill had gone through Texas hardpan like so much butter. It had never caused any problems, going in or coming out; it was, as Scott would say, "a nuthin.' " What Allen did not know—what no one knew—was the lunar soil had been so thoroughly compacted and tamped down by eons of micrometeorite rain that there was barely any room for the drill to penetrate. Once driven into the moon, it might as well have been in a vise.

"I don't think it's worth doing, Jim," Scott said. "We're not going to get it out." But Irwin wasn't ready to give up. He suggested that each of them hook an arm under one of the handles. That helped; they managed to pull the core about one-third of the way out. Now the men crouched down and each put a shoulder under a handle and tried to stand up: "One, two, *threeeee*—" More progress. Another push, then another. They were winning. Suddenly:

Helmsman in mission control for Apollo 15's moonwalks, flight director Gerry Griffin proved himself a valuable ally to mission scientists. When Scott and Irwin spent precious time trying to free the deep core sample from the moon's tenacious grip, Griffin kept impatient NASA managers from ordering them to abandon the effort.

Rover TV pictures show Scott, at left in the first image, and Irwin in their struggle with the core sample. Together, the men manage to withdraw the 10-foot-long core tube *(center)*. Then Scott carries it to the Rover for disassembly *(right)*.

"There we go!" The drill flew upward, and as Irwin said, "We almost flew with it!"

But Scott's troubles were not over. The entire core sample was 10 feet long; before they could bring the core home they would have to dismantle it into sections. And for some unknown reason the vise from the Rover's tool kit refused to work properly. Scott realized with exasperation that it had been assembled backward; no wonder it wouldn't grip. Irwin, meanwhile, broke out a hand wrench. Eleven more minutes went by while they struggled to dismantle the sections. Scott's patience was dwindling. Every minute spent on this one sample was time lost for the explorations to come. The trip to the North Complex hung in the balance. "How many hours do you want to spend on this drill, Joe?"

Scott wasn't the only one losing patience. For a while now, Allen could hear voices on the flight director's loop, from McDivitt and the others in the back row of mission control, pushing flight director Gerry Griffin to abandon the deep core and get on with the next item on the timeline. Irwin was to take a movie of Scott driving the Rover for the engineers. But Griffin, who was fully committed to the science of Apollo 15, wasn't going to be diverted. He understood the value of the deep core, and as the managers pressured him he walked over to Allen's console and said quietly, "You worry about that core. I'll take care of the back row."

On the moon, the struggle continued. Scott and Irwin managed to disassemble the core, but only partially; that would have to do for now. Twenty- eight minutes after they began, Allen told them to move on. They'd pick it up on the way back to the LM at the end of the traverse. Scott wasn't sure where they'd put it in the command module, but he'd think of something; that deep core cost too much to leave on the moon. When the deep core was finally scrutinized in the Lunar Receiving Lab, it would teach what geologist Don Wilhelms would call "the lesson of the moon's antiquity

and changelessness." Scientists would identify no less than forty-two separate layers of soil; the bottom layer had apparently remained undisturbed for half a billion years. ☾

5:17 A.M.

With the deep core struggle behind them, Scott and Irwin were glad to be heading west toward Hadley Rille. They had expected a short, easy drive, but instead found themselves pitching over dunelike ridges and troughs. Here, Irwin thought, was a place where they could get lost. But even as *Falcon* disappeared over the horizon behind him, the first time astronauts had ventured out of sight of their lander, Irwin felt no anxiety, so familiar was this valley to him now. Impatient to reach the rille, he urged Scott to take the shortest route, but Scott methodically stuck to the heading Allen had given them, telling Irwin, "We're making good time." As they neared the canyon's edge, Scott's attention was diverted to a small exploration he could not pass up, a small fresh crater that would prove the youngest ever visited on the moon, a scant million years old.

5:57 A.M.

When Scott and Irwin finally stood at the edge of Hadley Rille, they were rewarded with a sight no one had expected. On the far wall, which was in full sunlight, distinct layers of rock poked through a mantle of dust, like the levels of some ancient civilization. They were surely lava flows. Over many millions of years, perhaps, a succession of outpourings had piled atop one another to build up the valley floor. This was the first—and only—time that Apollo astronauts would find records of the moon's volcanic life, not as fragments scattered around the rim of an impact crater, but in place, preserved from the day they were formed. This was true lunar bedrock. And there was more of it on this side of the rille. Scott and Irwin gathered their tools and went to work.

Near the rim of Hadley Rille, Scott reaches for his hammer to knock fragments from a block of basalt. Running low on time and energy, he and Irwin gathered only a few samples at the rille before heading back to the lunar module.

There was no sharp dropoff at the rim of Hadley Rille; it was more like the gentle shoulder of a hill, and thankfully, the ground was firm. Effortlessly, Scott continued past the rim and loped several yards down the slope. Even now he could not see the bottom; it was hidden from view beyond the curved flank. He turned and ascended once more. All around him and Irwin were

Captured by the Rover's TV camera, Scott trips on a rock as he prepares to photograph the wall of Hadley Rille with a telephoto lens. In the moon's gentle gravity, a fall was merely an inconvenience, with almost no risk of injury.

big slabs of tan-colored basalt, shot through with holes from long-vacant gas bubbles; some of the boulders were scored by layers in miniature. But no sooner had Scott hammered off a chip than Joe Allen passed up word that he and Irwin would have to return to the Rover to collect a rake sample, and then it would be time to leave. On another day, Scott might have put up a fight; today he was too tired.

But in the back room, the geologists decided that the rille was worth more time. Urgently, Scott called Irwin to join him a little farther down the slope, where masses of darker rock waited. As they set to work, the men heard Joe Allen's voice once more.

"Out of sheer curiosity, how far back from . . . the edge of the rille are the two of you standing now?" Scott wasn't sure what he meant until Allen clarified it: "It looks like the two of you are standing on the edge of a precipice. . . ."

Scott managed to suppress a laugh. He could just imagine the back row of mission control—McDivitt, Kraft, Petrone, and the others, tied up in knots, their worst nightmare splashed on the big screen and multiplied on every TV monitor, because it looked like their boys on the moon were about to wander over the edge of a cliff. "Oh, gosh no," Scott reassured, "it slopes right on down here. . . ."

The extension amounted to all of eight minutes, enough time to gather a few more samples. Scott and Irwin headed back to the Rover wishing, once again, that they had more time. One thing was still true, and it would be true throughout the Apollo era: those on earth who carried the burden of responsibility for human lives would greet each new exploration with apprehension, while the men on the moon, confident and committed in their work, wanted only to push a little bit farther.

7:56 A.M.

Back at Hadley Base, Rover tracks converged on the foil-clad lunar module, and the ground was littered with gear and stowage bags attesting to the activity of the past three days. Beyond, the Apennines were bathed in sunlight, their slopes seemingly as soft and virginal as a meadow after a fresh snowfall. On the drive back, gazing at them, Jim Irwin had finally quoted the words of the Psalmist. But there had been no time for the North Complex; they would not solve that mystery after all. For all that they had accomplished, that loss was hard to take; Irwin felt their explorations were only half-finished. But right now, the most important thing was to get back into the LM with plenty of time to prepare for liftoff.

When everything was all packed up, Scott attended to a task that wasn't on the checklist. He had thought he might not get to it, but it was Joe Allen's idea, and he didn't want to disappoint him. He reached into a pocket on his suit and pulled out a falcon feather. Then, picking up his geology hammer, he loped back to where the Rover was parked and stood before the TV camera. ☾

"In my left hand, I have a feather. In my right hand, a hammer." Centuries before, the story was told, Galileo Galilei had stood atop the Leaning Tower of Pisa and dropped two weights of different sizes, proving that gravity acts equally on all objects regardless of their mass. And where would be a better place to confirm his findings, Scott said, than on the moon? He held the hammer and the feather out in front of him, and let go. They fell slowly through the vacuum, side by side, and for one brief moment the spirit of the great Italian scientist was conjured on the airless ground of Hadley Base. When the hammer and the feather hit the dust at the same time, there was applause in mission control. "Nothing like a little science on the moon," Scott said. ☾

8:08 A.M.

Alone, Scott drove Rover 1 to a small rise about 300 feet east of *Falcon.* From here, mission control would be able to aim the TV camera back at the LM and watch the liftoff, four hours from now. Scott made some final switch settings, and then, before he headed back, he pulled out a small red Bible and set it atop the Rover's control panel. If anyone should come this way again, he wanted them to understand who had left this machine here. Then he walked about twenty feet until he came to a small, subtle crater, reached down, and made a hollow in the dust. Into it he placed a small plaque bearing the names of fourteen men. The previous winter, at a New Year's party at Deke Slayton's house, Scott and Slayton had shared the hope that no more astronauts would ever be lost to the cause of space exploration. Scott resolved that he would find a way to memorialize those who had already been killed; maybe then, no

A plaque behind a small aluminum figure, which Scott commissioned from a New York sculptor, lists the names of 14 fallen astronauts and cosmonauts. Scott left the memorial on the moon in tribute to the intrepid pioneers who had died pursuing their nations' goals in space.

more would die. He had never met some of the men remembered on this plaque, including the three cosmonauts of Soyuz 11. Others he had called friends. Next to the plaque, he placed a small aluminum figure, a stylized representation of a fallen astronaut. ☾

Scott paused to take one last series of panoramic photos. Then, casting his eyes on the high slopes of a peak just to the south of Mount Hadley, he saw what looked like layers, and automatically began to describe them for the geologists. But Joe Allen cut him off; there wasn't time.

Scott took off in long, easy strides, heading across the rolling plains toward his lunar module. He could not deny a sense of loss, knowing that he would never return to this ultimate field site. But just now, Joe Allen was quoting the science fiction author Robert Heinlein: "We're ready for you to come again to the homes of men on the cool green hills of earth." Scott felt the chin strap of his communications carrier rasp against a week-old beard. He'd promised his children he wouldn't shave until he got home. After all: explorers always come home with a beard.

IV: THE FINAL SELECTION

The first thing Jim Irwin did when he climbed into the life raft was dip his hand into the cool water of the Pacific and bring it to his face. He wore no respirator. There was no quarantine trailer waiting for him and his crewmates on the carrier *Okinawa.* Three landing crews had returned to earth healthy; not a trace of anything alive had turned up in the samples, and at last NASA could do away with quarantine without risking any political heat. And Irwin was very glad of that. To arrive from the dust of Hadley to the blue Pacific was the most wonderful sensation he could have imagined. He wanted only to let the sea air wash over him and the morning sun fill him with warmth. He would have stayed there in that raft, but all too soon he heard the roar of the helicopter about fifty feet above him and felt its wind. The recovery people lowered the net for him; Irwin climbed into it and ascended from the water back into the sky.

Aboard the helicopter, Irwin tried to come up with something to say when he arrived on the carrier deck. Dave Scott reminded him and Worden that when they stepped out of the helicopter they should all salute in unison. They had talked about this before, Irwin remembered; Scott didn't like the fact that the other crews hadn't saluted together. On this flight, they were going to salute like military officers. Suddenly they were landing, and when

NASA

Scott *(left)*, Irwin, and Worden await recovery in a life raft after splashdown in the Pacific on August 7, 1971. The astronauts were the first Apollo crew not subjected to a 21-day quarantine upon returning from a visit to the surface of the moon.

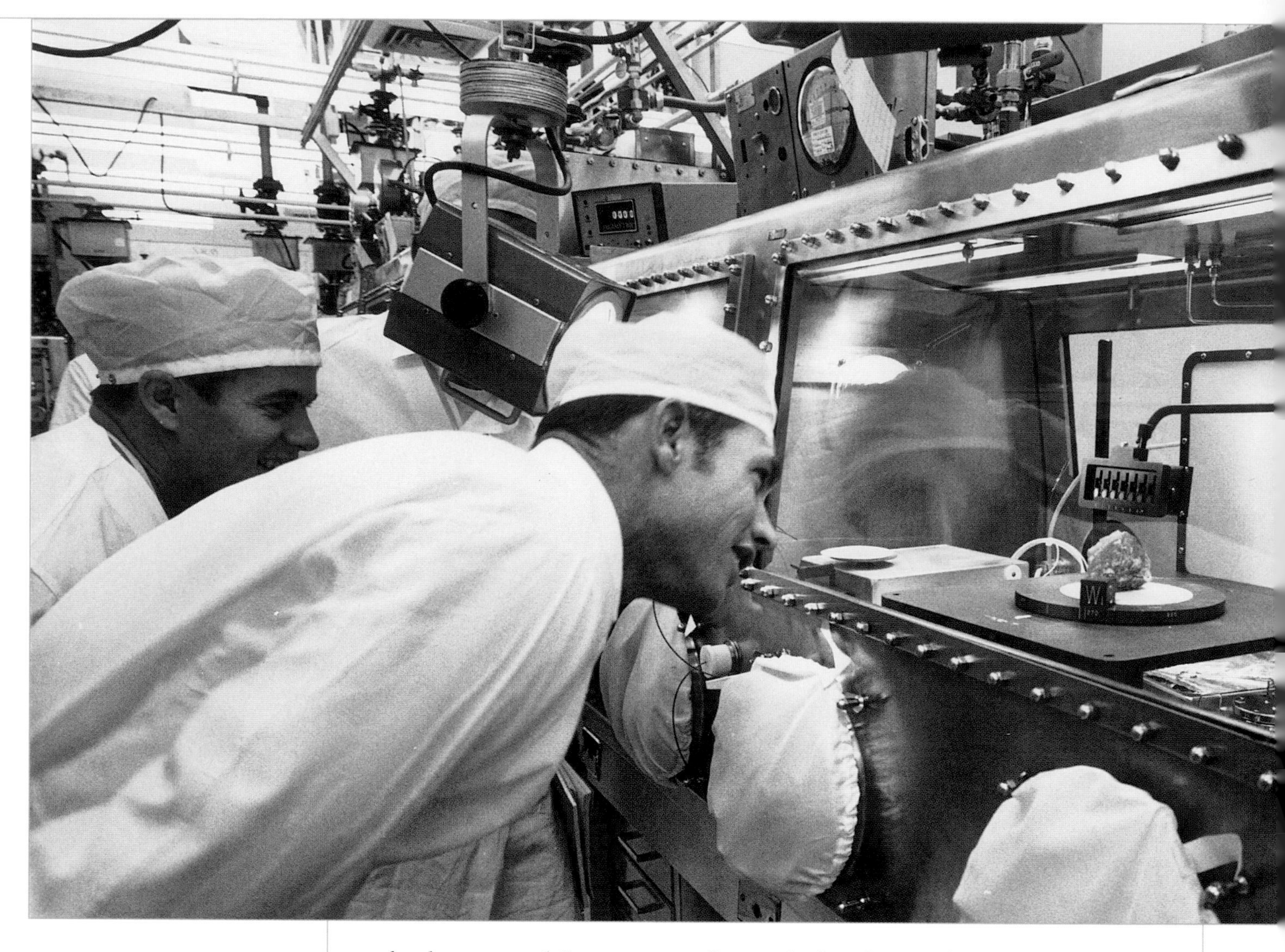

the door opened Scott was out first, and when he stood on the stairway he saluted immediately, without waiting for them.

On the carrier deck, Irwin felt unsteady and realized he hadn't cleared his ears in the command module on the way down, but those last five minutes had been the most hectic of the entire flight. One of the three parachutes had collapsed as soon as it was deployed, and the craft had descended faster than normal. On the way down Irwin had seen them have a near-miss with a helicopter. They'd hit the water with a solid impact, but everything was all right.

When it was his turn at the microphone, Irwin expressed gratitude for the trip, and to everyone on earth who supported them. He had been trained at the Naval Academy, he said, and he'd been on ships before, but he'd never been so glad to be on one as he was right now.

During the night Irwin slept fitfully; as he lay flat in his bunk he felt as if his body were tilted head-down at a steep angle. At five in the morning he awoke to the din of the anchor chains up on the deck. Not long afterward he

In the Lunar Receiving Laboratory on August 12 *(left)*, Scott, accompanied by Joe Allen, reacquaints himself with lunar sample 15415, otherwise known as the Genesis Rock. The white crystals seen in the sample's "mug shot" below are calcium feldspar, remnants of a primordial crust formed when the moon was very young—and very hot.

and his crewmates were getting the heroes' welcome at Hickam Air Force Base in Hawaii. By midmorning they were airborne again, inside a big transport headed for Texas. They were strangely isolated in their small compartment, as if everyone were avoiding them, as if they weren't sure ending the quarantine was such a good idea after all. Probably it was just as well, because no sooner had the C-140 lumbered off the runway at Hickam than Scott pulled out a stack of first-day covers for them to sign; he had carried them on the flight in the pocket of his space suit. The covers were part of a deal with a stamp dealer who had approached Scott during training; eventually, some of them would be sold in order to set up trust funds for their children. Before the flight, Irwin had harbored reservations about the deal, but had decided not to say anything to Scott for fear of causing friction with his commander. ☾

Irwin sat in silence, signing the envelopes and assembling in his mind the experiences of his twelve days in space. He thought of the wonderful mental clarity he'd had on the moon, the powerful sense of God's nearness. He thought of the discovery of the white rock. He could not imagine how his crewmates might have been affected, but here in this small cabin he knew his soul had been stirred. He was a nuts-and-bolts man who had come back with something he had never anticipated: the seed of spiritual awakening.

And when he was back home in Nassau Bay he found joy in the most ordinary things. Just to sit in a chair, to eat at a table without having to chase food into his mouth, the reassuring sense of up and down—he had a new appreciation for all these things. Life itself was cause for celebration. Irwin felt so renewed that he wondered if he'd discovered a cure for despondency, for those who have lost their appetite for life: lock them up for a while, deprive them of all earthly experience, then let them out into the world again.

But physically, Irwin wasn't quite right. He was still having difficulty adjusting to the pull of gravity, including some bouts of dizziness. He wasn't doing very well on the stress tests in the medical physicals. And the doctors had told him that during the flight, after he and Scott had returned to the command module, he'd had a problem with his heart. At the time, Irwin had no idea anything was wrong; he knew only that he was suddenly very tired.

Unbeknownst to him, the doctors watching his EKG in mission control were alarmed to see a so-called bigeminal rhythm, in which each normal heartbeat is followed by an extra beat. Houston never said anything about it, nor about the fact that both he and Scott had also experienced minor heart irregularities during their time on the lunar surface. Both men were angry that no one from Chuck Berry's staff had informed them while they were in space. Now, on earth, the doctors were ascribing the problems to the rigors of training. During practice sessions for the moonwalks, Scott and Irwin had spent hours working in space suits in intense summer heat. To replenish lost fluids, they drank large amounts of an electrolyte solution, which tended to leach potassium from their systems. As a result, the doctors would conclude, they left earth with a potassium deficiency that was exacerbated by the stress of the mission, causing their subsequent heart irregularities. Irwin hoped that soon he would be back to normal.

IN 1971

Ohio becomes the last state needed to ratify the 26th Amendment to the U.S. Constitution, lowering the voting age to 18.

The Carpenters win the Grammy for best pop group for "Close to You."

Walt Disney World opens near Orlando, Florida.

The *New York Times* begins publication of "The Pentagon Papers."

Short-shorts become fashionable as "hot pants."

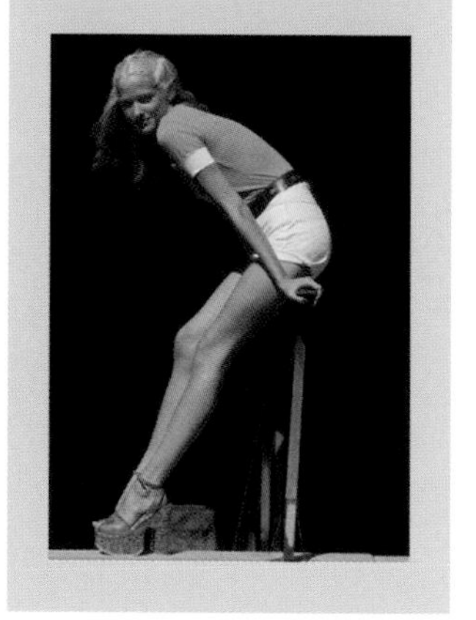

Dave Scott was also tired, though not as much as his lunar module pilot. He wished he and his crew could have been locked up along with the moon rocks in the Lunar Receiving Laboratory. He wanted nothing more than to be put away for three weeks, with time to rest and write the mission reports and talk to the scientists. He would have loved to have time to look at the videotapes of the moonwalks and study the rocks with geologists. There wouldn't have been any pressure from the home front to get him back quickly; it would have been fine with Lurton. Before the mission, he'd pushed hard for it, but not hard enough. Instead, he was let out into the world for long days of debriefings. In his free moments he found himself signing photographs; it seemed that every general in the air force wanted one autographed specially for him. And at night, it was the time of the neighborhood party. Scott would have enjoyed it much more if he hadn't been so worn out. And still to come, there was the trip to Washington, the speeches before Congress. Then to New York and Los Angeles for talk shows and parades. A month after splashdown, instead of savoring the accomplishment, he would be completely exhausted.

To Al Worden, the debriefings were interminable. Day after day they rehashed the mission for the engineers, and then the operations people, and then the managers. By the second week of answering the same questions over and over again Worden was coming home physically drained—sometimes arriving to a party in full swing in his apartment. He couldn't wind down until well past midnight. He was so overtired he couldn't sleep. Alone in his darkened living room, memories of the flight, and the training, came into his head, and he began to write about them. The words seemed to be coming from somewhere else. They flowed out of him. He began to jot down ideas, most not even complete sentences, page after page of them. It all came back—seeing the universe revealed above the far side of the moon, the battered moon itself, a tiny crescent earth adrift in blackness. And then there was the spacewalk. In the middle of the voyage back to earth, Worden had ventured outside *Endeavour* to retrieve cassettes of exposed film from the scientific cameras on the side of the service module, and he remembered every moment:

A spacewalk
Is like
Being let out
At night
For a swim
By Moby Dick.

It was as if there were another debriefing he had to perform on himself, to give his mental computer a chance to play back the images and impressions he had stored up in space. It was the data the engineers didn't ask about. Over the weeks he rewrote and rearranged the lines until they were poems, the poetry of a man caught in the lingering pull of another world:

Quietly, like a night bird, floating, soaring, wingless
We glide from shore to shore, curving and falling but not quite touching;
Earth: a distant memory seen in an instant of repose. . . .
I glide upward, above the waves of the ocean moon. She is forever moving just out of reach and I sail on,
never touching, only watching and wanting to know.

●◐◐○○○◑◑

In August 1971, NASA was on a high. With Apollo 15, the agency had scored its biggest success since the first lunar landing. Within the scientific community, even skeptics applauded the mission. Gerry Wasserburg, a Caltech geochemist who had voiced his share of opposition to Lee Silver's astronaut training efforts, publicly named it "One of the most brilliant missions in space science ever flown." For Dave Scott, there was an unaccustomed gesture

Dick Gordon *(left)*, training for the moon with his lunar module pilot Jack Schmitt, hoped that the two would be named to the final lunar landing mission, Apollo 17. Instead, Flight Crew Operations Director Deke Slayton named Schmitt to the crew for Apollo 17, but not Gordon—despite both men's entreaties to keep their team together.

of praise from a fellow astronaut: a note from Alan Bean congratulating him on a superb performance. And throughout the space center there was the new realization that Apollo had reached maturity. Almost every one of Apollo 15's innovations, from the Rover to the upgraded lunar module to the training and planning efforts, had come through with barely a hitch. The way was clear for the final two missions to be even more ambitious. ☾

And for six men, August was the moment of truth. The crew of Apollo 17, yet unnamed, was about to be announced. Six men eyed three seats on the final moon-bound command module. So did the scientific community. All through 1971 they had been turning up the heat on NASA, decrying the fact that NASA had a fully qualified scientist and wasn't sending him to the moon. Word of this filtered down to the Astronaut Office, where Dick Gordon had been making a hard run to snare Apollo 17 for his own. He knew very well that by Slayton's rotation it was Gene Cernan, who had backed up Shepard on 14, who was in line for the mission. When Cernan crashed his helicopter in the Banana River during training, Gordon had hoped Cernan had taken himself out of the running, but that little episode had just slid off Cernan's back. Now, as the decision neared, Gordon was betting that Schmitt would be his edge. And he had no qualms about politicking for the flight; he went to Shepard and Slayton and asked them to keep his crew together. ☾

From the sidelines, the other astronauts watched and waited with great interest. There were two ways this could go, they realized. Even if Gordon didn't get the mission, Slayton could still take Schmitt and put him on Cernan's crew—in which case things were looking very bad for Cernan's lunar module pilot, Joe Engle. He was one of the most experienced test pilots among the Original 19, and, in Cernan's mind, one of the best pilots he'd ever flown with.

But Engle had a little of Gordon Cooper's strap-it-on-and-go attitude, and it had already hurt him. During Apollo 14, he had taken his role as backup lunar module pilot too casually. The computer and all its software routines was daunting enough to any astronaut, but Engle just wasn't applying himself. Tom Stafford, now the chief astronaut, was coming down on Cernan—"You're the commander of this goddamn crew; get him in gear." And Cernan was telling the simulator instructors, "Whatever you can do to help Joe, do it." But ultimately it was up to Engle to change, and he didn't seem to realize how carefully his performance was being scrutinized.

But in truth, Deke Slayton wasn't especially concerned. In his mind, only one thing mattered; it would have been unthinkable for NASA to have a geologist-astronaut and not send him to the moon. ☾

Gene Cernan *(left)* and Ron Evans got the news of their selection for Apollo 17 while vacationing in Acapulco, Mexico. Before becoming astronauts, the two pilots had known each other at the Naval Postgraduate School in Monterey, California.

AUGUST 12

It was Thursday evening, and at the King's Inn Hotel on NASA Road 1, Jerry Schaber was listening to the evening news. Back in 1965, Jack Schmitt had hired Schaber to work at the USGS in Flagstaff; over the years they had become friends. Schaber had just come through one of the most exhilarating experiences he could remember, watching Scott and Irwin explore Hadley; nothing, he would say later, could compare with seeing your friends go to the moon. But on this Thursday evening Schaber's thoughts were on Jack Schmitt. The crew announcement was due any day now. Schaber was pessimistic; he couldn't imagine how Schmitt could get the flight when a popular pilot like Joe Engle was in line for it. He'd just bet his roommate Jim Head five dollars that Schmitt wouldn't make it.

Suddenly, on television the reporter was saying that the crew of Apollo 17 would include geologist-astronaut Jack Schmitt. Schaber yelled to Head, "Guess what they just said! Jack's on Apollo 17!" Schaber picked up the phone to call Schmitt. "Hey, congratulations," Schaber said, "we just heard!" The voice at the other end was characteristically gruff.

"It's not true! I haven't heard anything from NASA. Until NASA tells me, I don't know anything about it."

"Well, can we come over and help you wait for the call?"

"I don't care."

In the Nassau Bay Apartments across from the Manned Spacecraft Center, Jack Schmitt was sitting around, by himself. One bedroom was still full of unopened boxes he'd brought with him from the Survey in 1965.

"Gee, Jack," said Head, "it's awful quiet around here for a guy who's going to the moon."

"It's not true," Schmitt said.

Head went over to the bar, opened his briefcase, and pulled out five small airline whiskey bottles.

"What's that for?" asked Schmitt.

"That's in case you get the right call," said Head.

Suddenly the phone rang. It was Schmitt's sister, calling from Silver City to congratulate him. Schmitt said, "It's not true!"

More waiting, and then the phone rang again. Suddenly Schmitt got very quiet and businesslike. Schaber and Head looked at each other. "Yes, sir," they heard Schmitt say. "Yes, sir, I'll do the best job I can." Schmitt hung up the phone, went over to the bar, and very calmly unscrewed the caps from three of the little bottles of whiskey, picked them up together, and slugged them down. ☾

Schaber would always remember, with amazement, that no one from the space center called to congratulate Schmitt that night. He and Head ended up taking him out to a nearby Pizza Hut.

Meanwhile, in Acapulco, a vacationing Ron Evans was scuba diving with his daughter, Jaime, and when they came out of the water there was the rest of his family, and Gene Cernan and his family, all excited. Deke Slayton had just called with wonderful news. The children made a banner and strung it across the cabana; it said, "Apollo 17." But on both sides of the border there were pangs of sadness, as the ones who made it thought about the ones who didn't.

The next day, Jack Schmitt said to Dick Gordon, "Why don't you let me talk to Deke?"

"No, I've had my shot," Gordon said. "The decision's been made." Schmitt went to Slayton anyway and asked him not to break up the crew. He and Gordon had worked very well together, he said, they made a great team. But his effort was in vain; Slayton wasn't going to change his decision. In the days that followed, Schmitt also spoke with Joe Engle. It was an awkward time; Engle was visibly upset and Schmitt could tell he was bitter. But Engle handled himself as well as could be expected. He told a reporter that the toughest thing he could remember doing in a long time was explaining to his kids that he wasn't going to the moon. ☾

During a sampling stop near Hadley Rille, Dave Scott retrieves a Hasselblad camera and telephoto lens from the Rover. Beyond is the bright flank of Mount Hadley Delta.

Hadley Base is awash in sunlight at the end of Apollo 15's third moonwalk. In the back-ground are the hills of the Swann Range, which Scott and Irwin named for geologist Gordon Swann.

Late in Apollo 15's orbital tour of the moon, Worden photographed a gleaming crescent earth rising above the highlands of the lunar far side. Lighting conditions, combined with the peculiar reflective qualities of lunar soil, give the moon a chocolate brown hue.

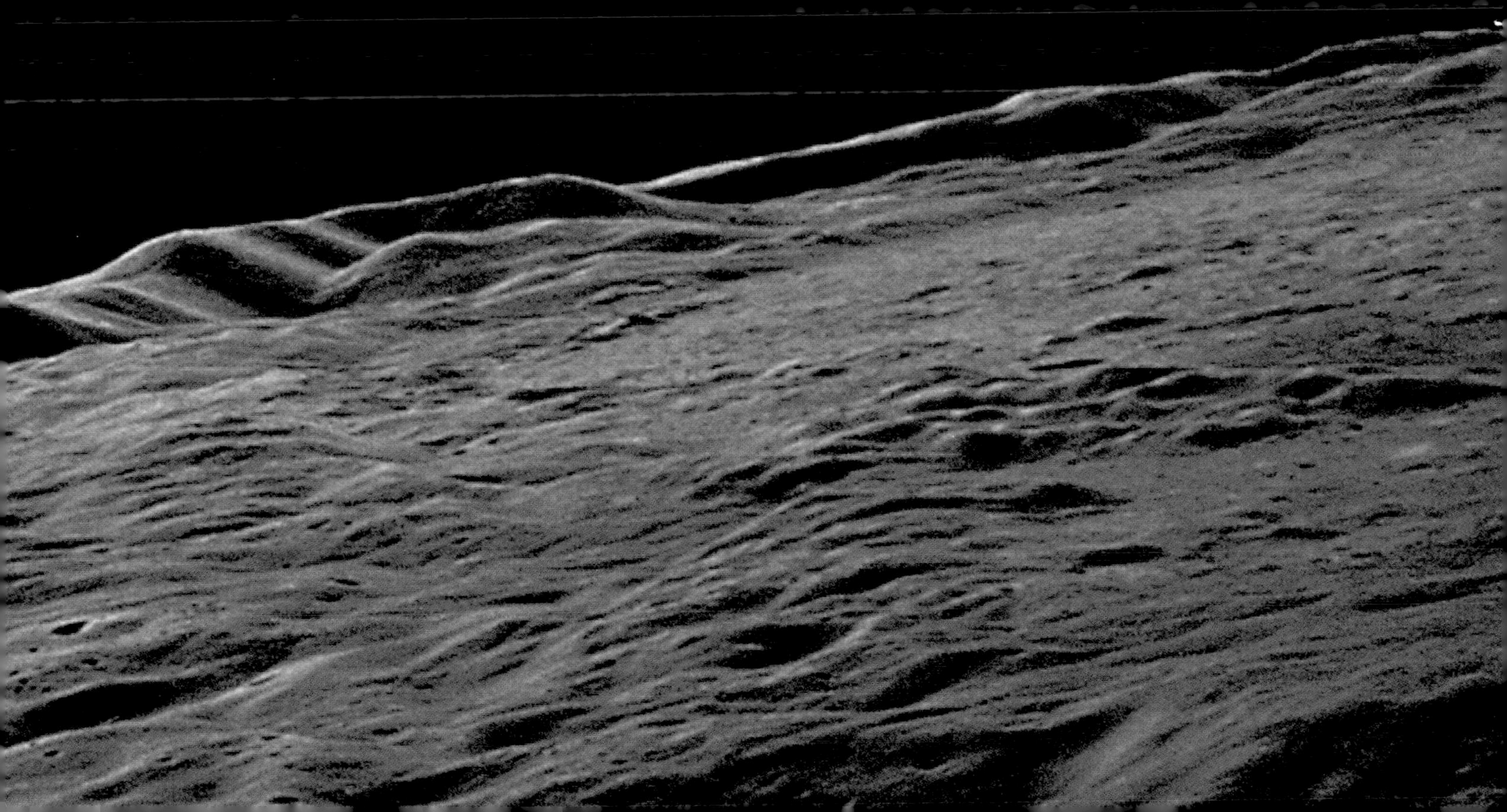

THE UNEXPECTED MOON

APOLLO 16

I: LUNA INCOGNITA

The moon that shone down on Houston in the spring of 1972 looked exactly the same as it had throughout human history. No telescope in existence could have picked up any sign that four teams of explorers had landed there, disturbed the stillness of the ages, and picked up rocks. But in the minds of the geologists who were analyzing those rocks (now totaling almost 385 pounds), the miles of photographic film taken on the surface and from orbit, and the rest of Apollo's burgeoning harvest of data, it was a world transformed. The moon of speculation was giving way to the moon of fact, and in many respects it was just as the geologists had theorized. The impact origin of most lunar craters, the formation of the *maria* by volcanic eruptions, and the existence of an anorthosite primordial crust had all been confirmed, though the details of these features would continue to be debated.

Apollo 16 commander John Young puts the Rover through its paces while Charlie Duke films the exercise. Young and Duke used their Rover to make the first explorations of the moon's central highlands in April 1972.

Knowing how lunar features formed was only one goal of the geologists; it was just as important to know when. With each new landing, the scientists had added more dates to their timeline of lunar history. Apollo 11 and 12 had given them the times of two different episodes of *mare*

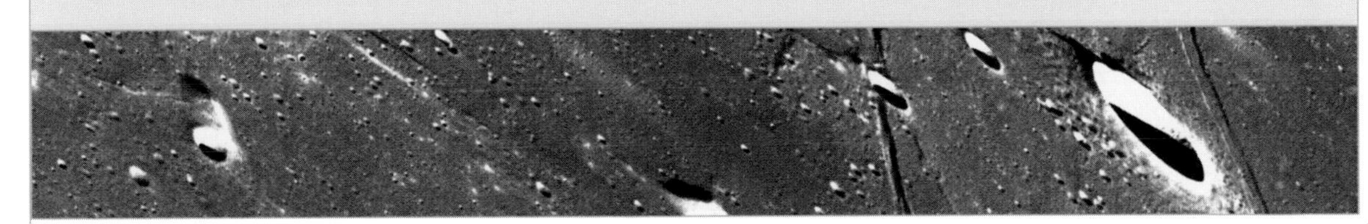

volcanism. And Apollo 14 gave them the best chance yet to look back before the *mare* eruptions to the moon's youth, and they saw an era of almost unimaginable violence. This was the period of heavy bombardment, when giant asteroids collided with the moon to form the impact basins. The bombardment came to an end when a chunk of debris the size of Rhode Island slammed into the moon, creating the Imbrium basin, 720 miles across, the largest and most spectacular crater on the lunar near side. Debris ejected from the impact appeared to have gouged, blanketed, or otherwise altered terrain for perhaps hundreds of miles around; the Fra Mauro hills were just one result. Even before Apollo, the geologists had realized that the date of that cataclysm—which reshaped so much of the near side—would be among the most important points on the timeline. With Apollo 14, they found it. The rocks from Cone crater gave a shaky but still usable date of 3.85 billion years. At Hadley Delta—which, along with the rest of the Apennines, formed the basin's rim—Scott and Irwin found rocks at Spur crater that confirmed this age. And thanks to the Genesis Rock, they could look back even further. Before Apollo no one had anticipated that the moon's primordial crust would be made of anorthosite, and many scientists still questioned the theory of a global magma ocean. But the idea was becoming harder and harder to refute; data from the sensors in Apollo 15's orbiting service module confirmed that anorthosite was a widespread component of the lunar highlands.

As the geologists were coming to understand the lunar surface, other scientists were lifting the veil of mystery from the interior. Including the ALSEP package at Hadley Base, there were now three fully equipped scientific stations scattered across the near side. Their seismometers were sending back data on tiny moonquakes caused by the pounding from small meteorites, and bigger ones triggered by the intentional crashes of spent lunar modules and Saturn boosters. For geophysicists, the shock waves from those quakes provided soundings of the moon's rocky depths. They revealed a crust some 40 miles thick, undoubtedly with variations from one region to another. Below the crust lies the lunar mantle, the source of the *mare* basalts. By analyzing those samples, the geochemists had determined that their source regions in the lunar mantle contained iron- and magnesium-rich rocks. By all signs, the mantle had cooled down enough to bring an end to the moon's volcanic activity eons ago. But was there a molten core farther down? Instruments had failed to detect a magnetic field, apart from traces of magnetic signatures preserved in some of the rocks. The earth's magnetic field is thought to arise from motions within a core made of hot, metallic fluid; by implication, the moon was probably fairly cool throughout.

With a titanic blast, an asteroid slams into the moon some 3.9 billion years ago to form the Imbrium basin, in this painting by Don Davis. An initial shock wave *(bright circle)* speeds away from the point of impact, where a huge volume of crushed and melted rock sprays outward. Immense impacts like this one were frequent events during the solar system's early history, on the earth and other planets as well as the moon.

The vivid colors in these moon rock photographs result from viewing a thin slice of the material through a polarizing microscope. Here a chunk of basalt from the Sea of Tranquillity, formed by volcanic eruption and collected during Apollo 11's moonwalk, reveals crystals of pyroxene *(yellow and purple)*, plagioclase feldspar *(white)*, and ilmenite *(black)*.

Similar externally to the moon rock above, this chunk of basalt was collected by Apollo 12 from the Ocean of Storms. Here, however, the pyroxene appears as needle-shaped crystals, indicating that this rock cooled more quickly than the Apollo 11 example.

Called an impact melt, this souvenir from Apollo 14's visit to the Fra Mauro highlands cooled from rock that was melted by the heat of a meteorite impact. The result is similar in texture to basalt.

Called a breccia, this Apollo 14 rock is an amalgam of rock fragments and smaller grains of various minerals, which were welded together in a glassy matrix *(black)* by the heat and pressure of a meteorite impact.

The celebrated Genesis Rock, collected during Apollo 15 *(pages 83-84)*, is a chunk of anorthosite, composed entirely of plagioclase feldspar. It is thought to have originated in a primordial ocean of molten rock that covered the moon in its infancy.

Geologists longed to send astronauts to Tycho crater, 51 miles in diameter, shown here in a mosaic of images taken by an unmanned Lunar Orbiter probe in 1967. Apollo managers, however, deemed Tycho too difficult to reach—and most of all, too rough for a safe landing.

This was where lunar research stood by the spring of 1972. There were many questions remaining, most notably about the composition and evolution of the highlands, the last great lunar unknown accessible to Apollo. Their lure was undeniable: The *mare* plains were the slate on which the last 3.6 billion years of lunar history was most clearly written, but only the battered highlands might contain a record of everything that came before. Sending astronauts to the ancient highlands had been one of the geologists' top priorities. The hills of Fra Mauro did not qualify, since their record of lunar history went no farther back than the Imbrium impact. To sample the oldest highlands, the astronauts would have to travel beyond the reach of Imbrium's extensive blanket of ejecta. One such landing site—one of the few that could be called a favorite of nearly all the geologists—was magnificent Tycho, the freshest large crater in the southern highlands and one of the most distinctive features on the lunar near side. A sharp unaided eye can spot Tycho from earth; its rays of ejecta stretch almost all the way across the full moon. On Lunar Orbiter photos giant boulders, torn from the depths of the highland crust, were visible along the rim of the 51-mile-diameter crater. Nearby, nestled among the hills, were ponds of once-molten rock that would make fine landing places for a lunar module. In January 1968, the final Surveyor probe had successfully touched down in one of these valleys. When it came time to choose a landing site for Apollo 16, the geologists pushed hard for Tycho. But their campaign hit a wall when NASA got a look at the hellishly rugged terrain that lay underneath the approach path to the landing site. Jim McDivitt, now the manager of the Apollo Spacecraft Program Office, saw the Lunar Orbiter photos and said flatly, "You will go to Tycho over my dead body." McDivitt's reluctance was compounded by the fact that Tycho was hundreds of miles away from the lunar equator. Getting there would take much more energy than previous missions; that would mean cutting down on Apollo 16's scientific payload. And more importantly, the trajectory necessary to reach Tycho was so far outside the free return that if the command module were disabled en route, an Apollo 13-type rescue would stretch the LM's fuel reserves uncomfortably far. Tycho was a risk NASA would not take.

The Descartes highlands were the scene of Apollo 16's three moonwalks, whose paths are shown here. Before the mission, some geologists argued that the shapes of the mountains supported the theory that they had formed by volcanic eruptions billions of years ago.

Instead, after some debate, the geologists chose a patch of hills west of the Sea of Nectar and not far from the 30-mile crater Descartes. Unlike Tycho, the Descartes highlands would probably not provide pieces of the ancient crust, which the geologists still hoped to sample on Apollo 17. But the geologists believed Descartes would yield an equally important prize, something no other place in the highlands seemed to offer: volcanic rocks.

The reason the geologists were willing to bank an entire mission on the promise of volcanic rocks was simple: they are the only direct means of knowing what has happened in a planet's interior. The *mare* basalts proved that the lunar mantle had once been partly molten, and that was tantamount to saying that the moon had been geologically alive. But those samples were only part of the story. The *maria* covered only 17 percent of the moon. Furthermore, the basalts brought back to earth had

Scientists brief the prime and backup crews for Apollo 16, aided by lunar photographs and maps. Clockwise from left: NASA geologist Fred Hörz, Charlie Duke, John Young, Stu Roosa, Ken Mattingly, and Ed Mitchell, partially hidden next to MIT professor Stanley Zisk.

formed during a fairly narrow, 350-million-year window of the moon's 4.6-billion-year history. If volcanic rocks could be found in the highlands, the geologists hoped, they might unlock greater expanses of lunar history before and after the *mare* eruptions. In short, they would tell the geologists how full the moon's geologic life had been.

Many of the mappers who studied pictures of the Descartes highlands saw ample evidence for volcanism. To these geologists, the shapes of the mountains suggested that they had been formed from lava, but one of different composition than found in the *maria.* On earth, basalt lavas are relatively thin and runny; they spread across the land like oil on a table top. Lavas whose composition is more like granite, on the other hand, are thick, like toothpaste. The difference is silica content; because granite has more silica, it is more viscous. Not only would such lavas produce landforms like those at Descartes, they would be lighter in color than basalts, just as the lunar highlands are lighter than the *maria.* When the geologists looked at the Descartes mountains—elongate mountains, with rounded tops and irregular shapes—they saw silica-rich volcanic rocks, which had never been found on the moon. Perhaps the mountains were made of rhyolite, the volcanic equivalent of granite. Or maybe they were piles of silica-rich volcanic ash and cinders.

Between the mountains, meanwhile, were patches of smooth plains that resembled other lunar regions that the geologists called the Cayley formation; they too were almost certainly volcanic deposits. They might even be the highlands' ancient version of *mare* basalt, brightened due to the effects of meteorite bombardment over the eons. In any case, Descartes seemed to offer what the geologists had long hoped for, a window into the evolution of the highlands. The scientists who had mapped the area believed some of its features might date back to almost 4 billion years, while others had formed only 1 billion years ago—fully two eons after the *mare* eruptions. Descartes was not only easy to reach, it appeared to lie beyond Imbrium's influence. For the geologists, Descartes was too good to pass up. And on April 16, 1972, when veteran moon voyager John Young and his crew—rookies Ken Mattingly and Charlie Duke—headed out of earth orbit, their primary goal was to make known Apollo's last *luna incognita.*

THURSDAY, APRIL 20, 1972
1:27 P.M., HOUSTON TIME
4 DAYS, 1 HOUR,
33 MINUTES MISSION ELAPSED TIME
ABOARD THE COMMAND MODULE *CASPER*

It's my fault. I've come all this way, and people have put all this time and money and effort into this, and I've managed to screw it up. Even as these thoughts flashed through Ken Mattingly's mind, he forced himself to focus on the job. Young and Duke were out there, drifting away from him, heading for their Powered Descent. And he was flying solo, getting ready to fire up his SPS engine and change *Casper*'s orbit from an ellipse with a high point of 69 miles and a low point of 9 miles—the so-called descent orbit—to a circle 69 miles above the moon. He'd routed electrical power to the engine, and turned on the gyros, and it was all normal, until he tested the secondary control system. There was a set of little thumbwheels that controlled the gimbal motors for the engine nozzle, and as soon as he touched the yaw thumbwheel, he felt *Casper* begin to shake. On the instrument panel in front of him, the 8-ball nodded back and forth. He pulled back his hand and the shaking stopped. He changed some switch settings and tried it again. More shaking. It felt like a train on a very bad track. He said aloud, "It's not gonna work." Once more, he switched settings and moved the thumbwheel. No change. He spoke to the empty cabin: "I be a sorry bird." ☾

Was it his fault? He *knew* he was doing everything just the way he'd done it a hundred times in the simulator. And yet, how could something have gone wrong with this beautiful flying machine that had worked perfectly for four days? Back at the Cape, a few weeks before the launch, he'd gone to the pad just to look at the Saturn V. And he'd realized, as he gazed at the towering

Shortly after beginning their moonward journey, the Apollo 16 astronauts had this view of their home planet from about 22,000 miles away. Much of the western United States is visible, and Baja California is particularly easy to spot.

machine, that he barely knew what he was looking at. Sure, he understood the basic design, and he knew the parts and pieces he had to know. But there were, what—several *million* parts in the whole thing? And each one had been designed, fabricated, tested, and installed by *someone.* Standing there, he knew the scope of Apollo was beyond the grasp of any one mind.

He rode the elevator up to the place where the third stage met the spacecraft adapter section, and there, at the juncture, was an open hatchway. He climbed through until he was standing inside a great metallic ring lined with pipes and electrical lines and all kinds of components. The lone technician who was working in there was startled—"Who are you? Get out of here."—but once he understood that he was talking to one of the men who would ride this rocket, he was just as gracious as could be. He said to Mattingly, "You know, I can't imagine what it's going to be like for you. But I can tell you this: It won't fail because of what *I* do." Mattingly realized that the reason Apollo worked at all was because thousands of people had said to themselves, "It won't fail because of me." From that moment on, Mattingly had taken that statement as his credo. He'd told himself as launch day grew near, and then, lying on his back in the command module, just before the engines lit, *It won't fail because of me.* How could he not feel the weight of the mission on him now.

"Hey, *Orion?*"

"Go ahead, Ken." John Young's voice.

"I have an unstable yaw gimbal number two . . ."

"Oh, boy." John Young, the last active member of the New 9, was the most experienced man in the Astronaut Office. He'd even been here before, the first man to make two trips into lunar orbit. He knew more about spaceflight than anyone Mattingly could think of. If only he were here right now, Mattingly thought, he might know how to work around this. But he was several thousand feet away. He couldn't see the engine, he couldn't feel *Casper* shake when he touched the gimbal control. To Mattingly went the honor of being the first astronaut to face a crisis alone, over the far side of the moon. He asked his commander, "You got any quick ideas?"

"No, I sure don't."

More than most astronauts, Mattingly thought, John Young seemed

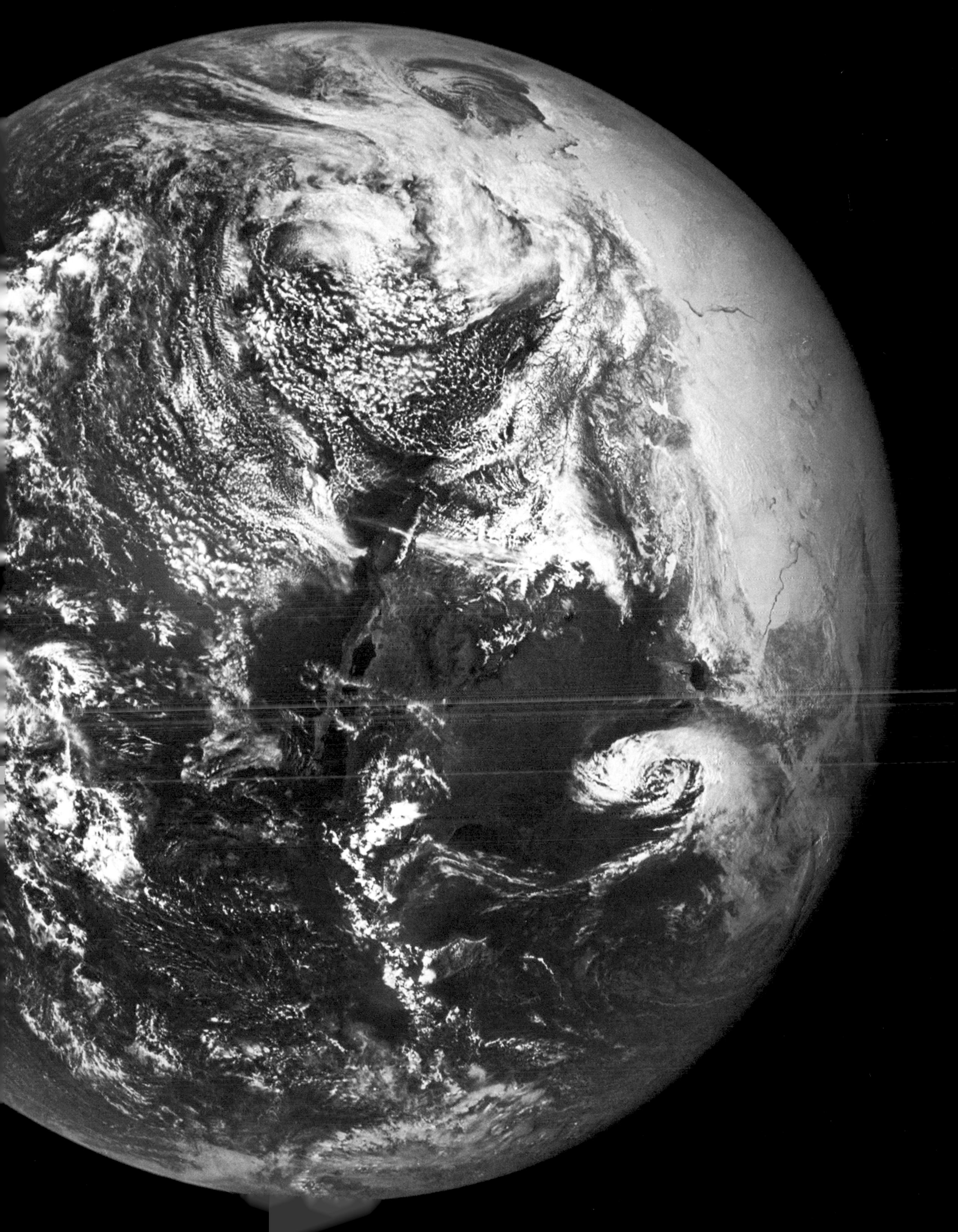

Concerned NASA officials hear from Apollo spacecraft program manager Jim McDivitt *(right)* about a problem with the backup controls for Apollo 16's service module engine—a glitch that could end the mission. From left: Rocco Petrone, John Holcolmb, Sig Sjoberg, Chet Lee, Chris Kraft *(seated)*, Dale Myers, and George Low.

mindful of the risks of his profession. Around the Astronaut Office, his memos were well known, sounding the alarm about some engineering problem he'd uncovered. He wouldn't rest until he knew every detail about the particular system or technique that worried him. And when he had learned all he could, then it was time to go fly—with his eyes wide open. That was the only way to handle this business; that was what made him so good. Maybe Young worried so much because he saw so clearly. But when it came down to the real question—Will you fly it?—John's answer would always be yes. And Mattingly could understand; he felt the same way. Maybe that's why he liked John Young so much.

Now Mattingly heard Charlie Duke's familiar drawl. "Hey, Ken, why don't you just stop it and start it again?"

"I've done that twice."

Mattingly knew his role as command module pilot. He was a truck driver: Get the big boys out to the moon, drop them off, and get them home. Secretly he'd always wished something would go wrong so he could show how good he was. Let the LM's ascent rocket give out before Young and

Duke could reach him; he'd go get them. Let the computer fail on the way back to earth—he'd get them home with a pencil and a few directions from mission control. But not this. Failure, Mattingly had always thought, was supposed to be clear-cut: it works or it doesn't. This was a gray area. Was the engine control really broken? The nozzle moved, after all; it just shook a lot. Could he still make a burn with it? He had no idea. And as much as he wanted to do something to save the mission, he knew that if he went ahead and tried to make the burn and the engine *didn't* work right, he would have made a very bad decision.

Now he heard Duke again. "What do your rules say, Ken?" Mattingly knew what the mission rules said; there'd been a change about a week before launch, and Mattingly was so mad about a change that late that he almost didn't read it. Now he wished he hadn't read it. It said that all four thrust-vector-control circuits, primary and secondary, had to be working or else the "circ" burn was forbidden. The only thing to do was rendezvous with *Orion,* come around to the near side, and get word from Houston.

Mission control agreed; Young and Duke would have to take a wave-off on the landing attempt, rendezvous with Mattingly, and fly in formation until there was some answer. That would be difficult; the onboard computers weren't programmed for the subtle nuances of orbital mechanics required to bring together a command module and lunar module that were flying in the same orbit, less than a mile apart. Mattingly would have to brute force it just point *Casper* at the lunar module and fire the thrusters. That would cost lots of maneuvering fuel, but there wasn't any other choice. ☾

The hours dragged on as *Casper* and *Orion* circled the moon in limbo, waiting for Houston's decision. Mattingly doubted they'd find a way out. If only they could have said, "You guys get back together; we'll figure this out and try again tomorrow." But that was out of the question; the sun was rising over the Descartes highlands, and by tomorrow the lighting conditions would be no good for a landing. They would have to solve this today or call off the mission.

"*Casper,* we'd like you to . . ." Jim Irwin was telling Mattingly to try the gimbal motors again; mission control would record the telemetry and show it to the SPS experts. He could picture what was going on in Houston right now—Apollo spacecraft program manager Jim McDivitt overseeing the troubleshooting effort; phone lines to Downey and to MIT; engineers poring over data. And he knew, after witnessing Apollo 13, that given enough time, the experts on earth could solve any problem. But they couldn't possibly have enough time on this one. Mattingly was sure the mission was over. And he could tell by the dejected tone in Young and Duke's voices that they thought so too.

The funny thing was, Mattingly had a feeling he knew what might be wrong. A long time ago, he'd heard about a test of the engine in which the cable that carried signals to and from the gimbal motors was a little too short; when the nozzle was pointed to one side the cable went taut and pulled some pins out of its electrical connector. Mattingly wondered if that's what had happened to the secondary control system. And if so, then how could he be sure the primary system was okay? All the control signals went through the same cable. No point in worrying Young and Duke about this; he kept the thought to himself. As *Casper* and *Orion* circled the moon, Mattingly wondered just how much of his spacecraft really worked. He knew that if they had to, he and his crewmates could use the lander's engine to get out of lunar orbit. Was that what they were getting ready to do? Was this mission really over?

MISSION OPERATIONS CONTROL ROOM

In the back row of mission control, Jim McDivitt waited for *Casper* and *Orion* to come back around from the far side of the moon for the fifteenth time. McDivitt had been the manager of the Apollo Spacecraft Program Office for three years now, having taken the job after coming home from Apollo 9. Often, during the missions, he had decided whether or not to continue after emergencies struck. On Apollo 12, it had been his recommendation to leave earth orbit after the lightning strike. Apollo 13 had been his ordeal as much as anyone's at NASA. On the flight Al Shepard had thought might be trouble-free, McDivitt had faced Apollo 14's problems with a balky docking mechanism, and then, hours away from Powered Descent, an errant abort signal. Now it was McDivitt's turn again, as Apollo 16 hung in the balance.

There were some astronauts who would keep flying as long as you let them—no one more than John Young—but not McDivitt. There were only four ways to get out of this business: Get grounded, get fired, get killed, or retire. In McDivitt's mind, the only good choice was the last one, but if he waited too long, it would be one of the other three. After commanding Apollo 9 —which had been enough of a test for any astronaut—he knew he would probably command one of the later lunar landings, but he had never viewed going to the moon as an end in itself. He would have stayed to fly the first landing, but not the fourth. He had always wanted to manage a big program, preferably in government, where there would be more control. When the

IN 1972

The 11th Winter Olympic Games are held in Sapporo, Japan.

A coal-slag dam in Buffalo Creek, West Virginia, collapses after torrential rains, killing 118 people.

U.S. mining of North Vietnamese ports begins with Haiphong harbor.

The British government imposes direct rule on Northern Ireland.

***Tapestry* wins best album and best female pop vocalist Grammys for Carole King.**

program office wanted him, he felt it would be the best use of his abilities.

Now, three years later, he had no regrets. But the job had taxed him in ways that flying never had. Tracking lunar modules and command modules as they went through manufacturing and testing, shipment to the Cape, and final checkout for launch, and making sure everything happened on schedule to meet each new launch window, was as complex as anything he'd ever done. He could fully appreciate now the awesome responsibilities faced by Kraft, Petrone, and the other Apollo managers.

He'd never forgotten that going to the moon was dangerous. But in his mind, NASA accepted that fact the moment the Saturn V left the pad. With each successive step—leaving earth orbit, getting out to the moon, going into lunar orbit, and finally landing—the danger increased, but so did the *investment,* in terms of risk, to get there. Now that Young's crew were circling the moon, McDivitt was prepared—as long as they were safe—to fight even harder to keep them going than he would have if they'd been in earth orbit. A short time ago McDivitt had looked at the strip-charts of the telemetry from *Casper,* and it appeared that despite the oscillations, steering signals were getting through to the engine. At Downey, engineers had been analyzing the data and feeding it into a mockup of the SPS. They concluded that if Mattingly had to use the secondary system, the engine might shake, but it would be controllable. Was their data enough? McDivitt needed an extra measure of confidence, and as he had many times in this job, he got it from his own experiences on Apollo 9. In one of the many what-if exercises of that engineering test flight, they'd caused the SPS engine to shake while it was being fired, and it had still performed well. Just now, while *Casper* and *Orion* were behind the moon, McDivitt had met with Kraft and the other managers and told them it was safe to proceed. And once more, having taken the risk of sending men to the moon, NASA was about to make good on its investment.

5:55 P.M.

Mattingly couldn't believe it when Irwin called up and said they'd looked at the test data and figured out that everything would be okay. Whoever was responsible for this spectacular save, Mattingly was going to buy them a case of beer. But didn't they realize that the primary system might be out too? Well, if their judgment said go ahead, he wasn't about to argue, not after six years of working his tail off to get here. He marveled at the boldness that seemed to have filled the managers after more than a decade in this business.

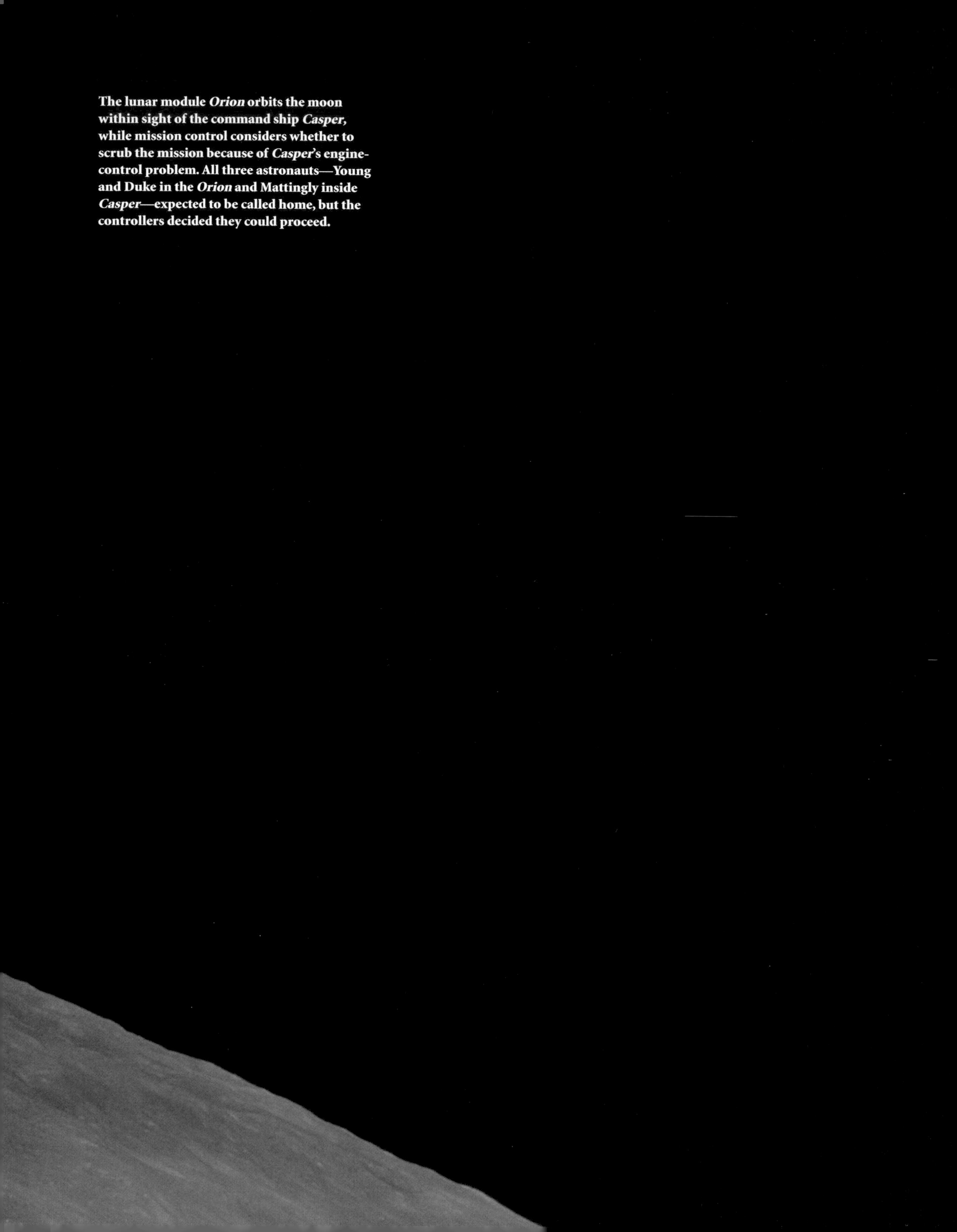

The lunar module *Orion* orbits the moon within sight of the command ship *Casper,* while mission control considers whether to scrub the mission because of *Casper's* engine-control problem. All three astronauts—Young and Duke in the *Orion* and Mattingly inside *Casper*—expected to be called home, but the controllers decided they could proceed.

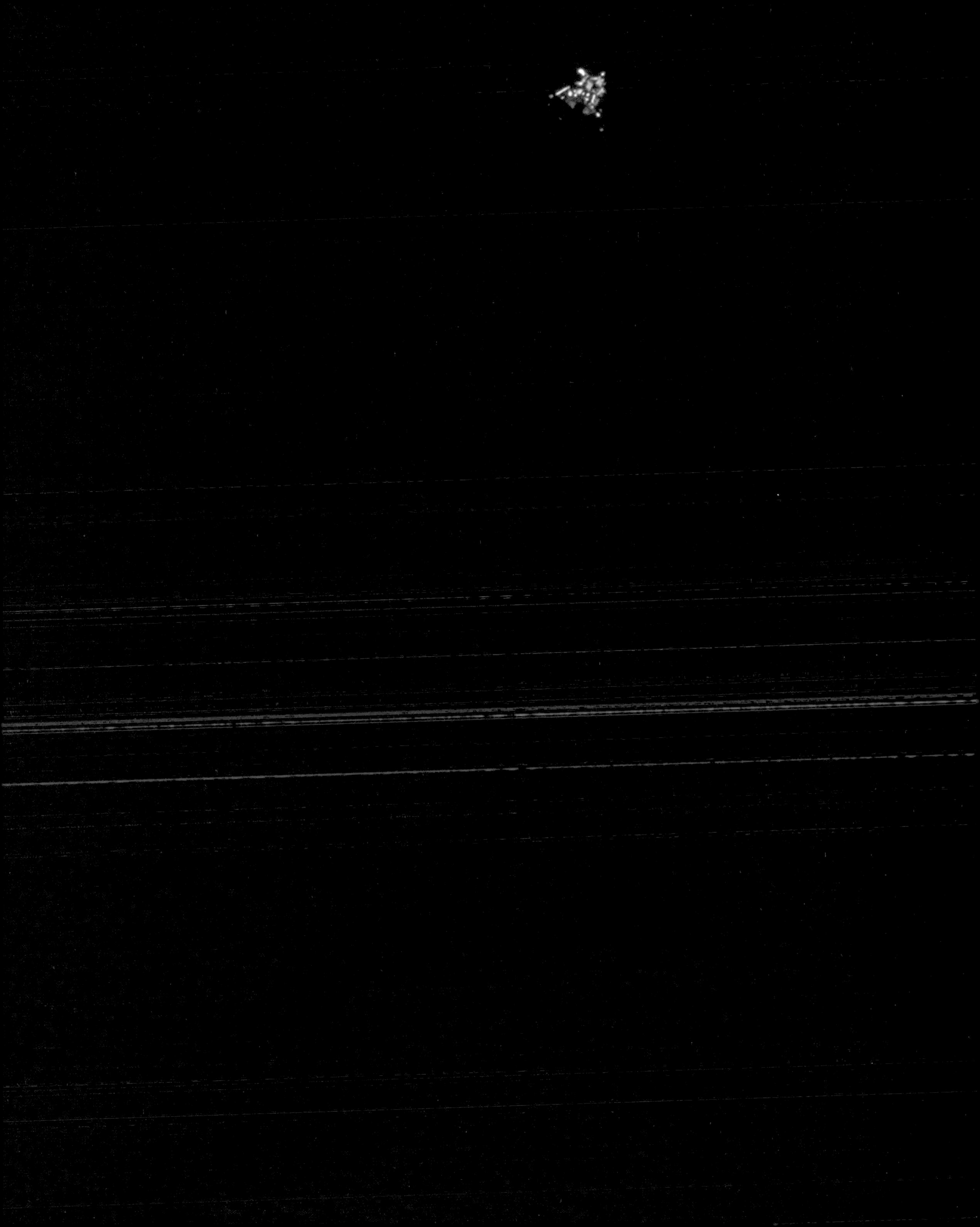

Orion casts a shadow on the Descartes highlands as Young steers the lunar module the last few feet to a landing. Clearly visible are the shapes of the craft's descent stage, landing legs, and descent engine nozzle.

When he was back on earth, Mattingly would tell McDivitt that he couldn't believe he let the mission continue, not when both of those control lines went through the same cable. And McDivitt would tease, "We *didn't* know they went through the same cable—you're the only one who did! You're right; we *wouldn't* have let you land!" ☾

8:23 P.M.
IN THE LUNAR MODULE *ORION*

"Okay, eighty feet, down at three. Looking super. There's dust." *Orion* sank straight down toward the Descartes highlands, kicking up wisps of bright soil. John Young kept his gaze out the window while Charlie Duke gave him data. "Let her on down. Six percent; plenty fat." If the past several hours had been rough on Ken Mattingly, they'd been even more traumatic for Young and Duke. Before undocking, *Orion* had an antenna problem, and then, because of a failed regulator, a dangerous buildup of pressure in the tanks of maneuvering fuel that threatened to abort the flight until Houston suggested shunting propellant into the ascent engine tanks. And then came the wave-off. They circled the moon, feeling their mission slipping away from them. But now, six hours behind schedule, Young and Duke were going to have their landing. *Orion* was flying just like the LLTV at Ellington, and after edging past one last

50-foot crater, the lander thumped to a stop. Charlie Duke, never one to restrain his excitement, couldn't contain himself. "Wowwww! Whoa, man! . . . Old *Orion is* finally here, Houston! Fan*tas*tic!"

"Well," drawled Young, "we don't have to walk far to pick up rocks, Houston. We're among 'em."

FRIDAY, APRIL 21
10:57 A.M., HOUSTON TIME
4 DAYS, 23 HOURS, 3 MINUTES MISSION ELAPSED TIME
DESCARTES HIGHLANDS

"Hey, John, hurry up!"

"I'm hurrying," said John Young as he made his way down *Orion*'s ladder. And then, for a moment, as he stood at last on lunar ground, he raised both fists in triumph. "There you are, our mysterious and unknown Descartes highland plains. Apollo 16 is gonna *change your image.*" No one listening on earth knew how prophetic those words would turn out to be.

Young added, "I'm glad they got ol' Brer Rabbit, here, back in the briar patch where he belongs." Years later, Young would be reluctant to explain the quote, but it seems clear that Brer Rabbit is himself, and the briar patch is spaceflight. Even now, on his fourth space mission, he showed no signs of wanting to do anything else. But Young hardly looked the part of a seasoned spacecraft commander. One day a simulator instructor was showing a friend

Recorded by *Orion*'s onboard movie camera, Young raises his arms in a gesture of triumph as he takes his first steps on the Descartes highlands. Apollo 16 was Young's fourth spaceflight, and he was enjoying this one as much as the first.

John Young takes a breather during a geology trip to Canada in July 1971. As commander of Apollo 16, Young became the first man to journey twice into lunar orbit.

around the space center, and when he pointed out Young across the room—this rumpled little guy in a Ban-Lon shirt and jeans—the reaction was, "*He's* an astronaut?" Even people who worked for the NASA contractors didn't know what to make of him. He drawled his way through conversation and gave the impression he was still the quiet country boy who grew up in Orlando, Florida, back when it was mostly farmland. You could see them sizing him up: *This poor guy just isn't on top of it—I'll have to go slow here*—but before the day was over they learned differently. Young would sit through a presentation on the computer software without saying a word, and when the specialist had finished, he'd drawl, "Well, I don't know about the W-matrix or anything like that, but what gets me is, how come if that's true, when I do . . . this, I get . . . *this?*" And you could almost hear the floorboards falling away in the engineer's mind, as he thought it through: Hey, he's *right!* Some people saw the country-boy bit as an act; it wasn't. It was just John's way of getting the people around him to think a little harder about the problem. In any case, you only made the mistake once: John Young was no hayseed. Ken Mattingly would call Young one of the best-read people he had ever met. His sharp, intuitive approach to engineering problems was well known to his colleagues. Inside Young was an unwavering determination, an overriding sense of responsibility—to the space program, to the country, to his crew—and an almost childlike sense of wonder at the universe.

When it came to lunar field geology, Young had caught the spark from Lee Silver, beginning on that first trip to the Orocopias with Lovell and Haise, and he hadn't lost it. At that time, he and Charlie Duke had been looking forward to Apollo 16 as the first J-mission. In 1970 the pair were doing practice geology work at the USGS training site in Flagstaff when they found out they would lose that milestone to Dave Scott's crew; it was a tremendous disappointment. Now they were faced with the task of equaling or even surpassing Apollo 15's scientific haul, and Young was ready. "Oh, look at all those beautiful *rocks!*"

SCIENCE OPERATIONS ROOM MANNED SPACECRAFT CENTER

In the geology back room, William Muehlberger inspected a photomap of the Descartes highlands, ready to oversee his surface geology team. He was new at this game, and he'd had a very stressful twenty-four hours. The geologists had done their share of sweating during the wave-off. Then, during the night, they'd learned that because of the late landing, the managers wanted to cancel the third moonwalk, and Muehlberger had set his planning team to work on a position paper to convince them to keep it. They were up all night, and in the morning they made their case. The managers were persuaded, but the go-ahead would depend on whether Young and Duke could stretch their supplies of power, oxygen, and cooling water to allow it. Now, beneath his excitement that Young and Duke were finally about to face the mysteries of the Descartes highlands, Muehlberger worried that they might not have time to unravel them.

Muehlberger's tall and powerful form recalled his days on the football team at Caltech, where he had been a classmate of Lee Silver's. Unlike the disheveled, lived-in look of some geologists, Muehlberger was a picture of neatness; he was almost dapper. A professor at the University of Texas, Muehlberger had joined the Apollo geology effort after Apollo 14. For Apollo 16, he had taken over not only as chief of the Surface Geology Team, but for Young and Duke's geology training; Lee Silver had been too busy with his work on Apollo 15, not to mention his teaching load at Caltech. Now, seated at the back of the Science Operations Room, Muehlberger listened to the voices of his former students coming from the moon.

Since the preparations for Apollo 16 began Muehlberger had heard more about the volcanic interpretation of the Descartes highlands than he could remember. One of the unique aspects of Apollo's lunar explorations was that the geologists had to predict what the astronauts would find at each landing site long before the flight began. That wasn't the way Muehlberger or most other geologists were used to doing business, but NASA wanted the geologic objectives to be as carefully planned as any other aspect of the missions. And it would have been out of the question to convey any uncertainty to the managers. Had he or his colleagues stood up at a site selection meeting and said, "We're really not sure what we'll find," they would have weakened their position, and possibly even lost the chance to go to Descartes. ☾

Just as on previous missions, the geologists had made their predictions about Descartes using the discipline of photogeology. In this effort, their basic tools were their eyes, their minds, and photographs of the moon, along with knowledge gained from unmanned probes and the previous Apollo landings. Muehlberger knew that there were scientists in other fields who had nothing but disdain for photogeology. To them, any research that based its conclusions on photographs, or even worse, visual observations, was second-rate at best. Even some geologists agreed with them. But neither they nor anyone else could offer an alternative; there wasn't one, save going there. And each new mission seemed to confirm photogeology's predictive powers. The last four landings had given their share of surprises, but they had also validated a lot of good, solid scientific effort.

But with Descartes, the geologists were at a disadvantage. The best

"Hot dawg, is this great!
That first step on the lunar surface is super!"
—Charlie Duke

pictures available were the ones Stu Roosa took when Apollo 14's Hycon camera failed. For all of Roosa's skill, the smallest details they showed were calculated at 66 feet across. The challenge of analyzing Roosa's pictures went to USGS geologists Don Elston and Gene Boudette, who subjected the pictures to the most intensive scrutiny imaginable. Magnifying them under stereographic plotters, Elston and Boudette charted details that weren't much bigger than the grains in the film, including boulders they said were about 16 feet across. After a heroic effort, their finished map was dotted with all kinds of volcanic features, from lava flows to cinder cones to explosion craters. Muehlberger was amazed. Sometimes, when he looked into Elston and Boudette's stereo plotters, he wondered why he couldn't see the tiny details they talked about. But he knew Elston and Boudette had been at this for years, and they seemed sure of their conclusions—in fact, out of all the geologists, they were the most enthusiastic about the volcanic hypothesis. Muehlberger had had second thoughts about that scenario; after all, some experts had expected volcanic rocks to turn up at Fra Mauro, and none did. But he never really questioned the interpretation. And if others on his team harbored doubts about volcanics in the Descartes highlands, they hadn't voiced them loud enough for him to hear.

IN 1972

Arthur Bremer shoots Alabama governor George Wallace in Laurel, Maryland.

The Supreme Court finds the death penalty to be "cruel and unusual punishment," violating the Eighth Amendment to the Constitution.

Sandy Koufax and Yogi Berra are inducted into the Baseball Hall of Fame.

President Nixon meets with Mao Tse-tung during the first-ever official visit to China by an American president.

The late landing had thrown everything into disarray. Already, they'd lost the chance to have Young and Duke take telephoto pictures of Stone Mountain in the early morning light, to find out whether it had a pattern of linear grooves like Mount Hadley, and whether it might be an illusion caused by lighting conditions. But today's excursion was packed with objectives, and as Muehlberger listened to the voices of his two students, he heard the most exuberant pair yet to reach the surface of the moon.

"Hot *dawg*, is this great! That first step on the lunar surface is *super!*" No surprise that Charlie Duke should be so excited; he was often that way, like an overgrown kid. On the field trips Muehlberger had heard him chattering away in his South Carolina drawl while John Young mumbled in the background in a steady stream of one-liners. The grind of training seemed to bring out the vaudeville in them. Now, on the moon, it was the John and Charlie Show all over again. Presently, Duke was doing his impression of W. C. Fields: "Here's the ol' photomap. . . . Just like training. . . . A picture of Hadley Rille. . . ."

Young, meanwhile, was setting up a telescopic camera to photograph ultraviolet radiation from stars, and was discovering how much fun it was to handle the instrument, which was heavy and cumbersome on earth, in lunar gravity. Young's normal reticence was gone.

"Look at that, Charlie! Look at me carry it! I'm carrying it over my *shoulder!* Ha ha ha!"

Soon the Rover's TV camera was on, and everyone on earth could see the pair in action. They had planted the American flag, and now Duke was composing a picture of his commander. "Hey, John, this is perfect, with the LM and the Rover and you and Stone Mountain. And the old Flag. Come on out here and give me a big navy salute." Young did just that—and he did it while jumping three feet off the ground.

1:07 P.M.
SCIENCE OPERATIONS ROOM

Mark Langseth, a tall man in his thirties with brown hair, wire-rimmed glasses, and long sideburns, had taken a seat next to Jim Lovell to watch Charlie Duke drill into the moon. It was customary, during an ALSEP deployment, for a Principal Investigator to take this seat while watching his experiment being deployed. Langseth, a geophysicist at Columbia University, had been working for six years to find out how much heat was flowing from the lunar

An exuberant Young salutes the flag while leaping three feet off the ground, as photographed by Charlie Duke—and televised by the Rover's camera *(inset, above).*

As Young walks among Apollo 16's scientific experiments, his boot snags a cable *(top)*. Realizing what's happened, he kneels to examine the damage *(middle)*, then stands up holding the severed cable *(bottom)*. The misstep, which ruined Apollo 16's heat flow experiment, was televised as it happened by the Rover's camera.

interior. His experiment had been aboard Apollo 13; it burned up, along with *Aquarius,* in the atmosphere. At Hadley Base, Dave Scott's best efforts hadn't succeeded in getting the thermometers down to the desired 10-foot depth; Langseth got his readings, but the temperatures were twice as high as he expected. Only new data from another location would let him interpret the Apollo 15 results. Langseth was especially eager to see whether the heat flow would be different in the highlands. And it looked as if he was going to get what he wanted. Duke had the benefit of a redesigned drill, and even the moon seemed to be cooperating.

"Look at that beauty go!" Duke raved. Within minutes the hole was drilled and the thermometer inserted to its full depth. Duke beamed, "Mark has his first one all the way in to the red mark on the Cayley Plains." Langseth beamed too.

"Outstanding," said Capcom Tony England. "The first one in the highlands."

To which John Young quipped, "Ask him what we're going to do if the temperature shows like it does at Hadley." Young was working on the ALSEP, standing among the sensitive instruments and their ribbons of electronic cable. The camera was pointed at him now, as he started to walk away from the Central Station. Langseth saw something wrong. He nudged Lovell and said, "Look, he's got the cable around his foot." Before Lovell could call mission control, everyone saw it pull free of the Central Station. Judging from the cable's position, Langseth thought it must belong to the seismometer. But that thought didn't last long.

"Charlie?" Young's voice was almost plaintive.

"What?"

"Something happened here."

"What happened?"

"I don't know. Here's a line that pulled loose."

Duke, who had been drilling the hole for the second heat-flow probe, stopped and looked toward Young. "Uh-oh."

Young sounded worried. "What is that? What line is it?"

"That's the heat flow. You've pulled it off." Langseth couldn't believe what he was hearing. Neither could John Young.

"I don't know how it happened. God almighty."

"Well," Duke said flatly, "I'm wasting my time."

"God, I'm sorry. I didn't—I didn't even know it." You could hear it in Young's voice; he really had no idea it was happening. Like any moonwalker, Young could not easily see his own feet as he walked; for one thing, the small chest-mounted control unit was in his way, and also, he had to lean unnaturally far forward to compensate for the mass of his backpack. And it was well known that the ALSEP cables developed a "memory" from being rolled up, and once unrolled they refused to lay flat in one-sixth g. That was something Young and Duke had noticed in training, and they'd warned the engineers about it. But that didn't matter now. Cocooned within his suit, Young never felt the ribbon snag on his boot, pull taut, and sever, crippling Langseth's experiment.

"Tell Mark we're sorry," Duke said. "Is there no way we can recover from that, Tony?"

"I'm sure we're working it," England said. It was true; even now, as a stunned Langseth absorbed what had happened, experts were coming together to see if it was realistic to try a repair.

3:04 P.M.
SCIENCE OPERATIONS ROOM

Bill Muehlberger studied the photomap, tracing Young and Duke's progress as they drove west on their first geology traverse. With only a couple of hours available, the men would just have time to drive to some relatively nearby craters and get their first samples of the Cayley Plains.

For months, Muehlberger and his team had speculated on what the Cayley Plains would be like. Today, in TV pictures that were even clearer than Apollo 15's transmissions, they were seeing it. "Old *Orion*" had set down in a small "inlet" between two mountains that were aligned roughly north-south, like the peninsulas of San Francisco. On the monitors in the back room, the place did not appear to be very different from Hadley Base. There was the same expanse of undulating, crater-pocked, gray desert. And in the background, there were mountains, not as distinctive a skyline as the Apennines, but similar in their rounded form. But if the geologists' ideas were correct, the similarities were deceptive; those mountains, like the plains, were made of volcanic rock.

But so far, Young and Duke hadn't found what everyone expected. The

At Plum crater, Young, geology hammer in hand, pretends to sneak up on a boulder before chipping off a sample. Geologists had told Young and Duke to expect volcanic rocks at Plum crater, but this one—and the others they picked up at this site—turned out to be breccias *(pages 120-121).*

first rock Duke picked up, back at the ALSEP site, had not been a piece of rhyolite, or basalt, or any other volcanic rock. It had been a breccia—an impact rock. A breccia is a mixture of rock fragments and soil particles welded together by the enormous energy of a meteorite impact. Ever since Gene Shoemaker had found breccias at Meteor crater, geologists had expected to find them on the moon. Sure enough, minor amounts of these rocks turned up in the *maria,* and breccias had dominated the samples brought back from Fra Mauro. Some were universes in miniature, containing pieces of a host of separate rocks, occasionally even fragments of preexisting breccias. The information locked in one such sample can keep a geochemist, armed with an electron microprobe, busy for years.

Tony England, Capcom during Apollo 16's moonwalks, served as Young and Duke's link with the geologists in the back room. England had earned a Ph.D. in geophysics before becoming an astronaut in 1967.

Now, as the men drove west, Muehlberger could hear Duke describing some nearby boulders; he said he thought they were breccias too. Via Tony England, Muehlberger sent up a question: "Have you seen any rocks that you're certain aren't breccias?"

"Negative," Duke answered.

Well, some breccias were to be expected anywhere there were impact craters. And judging from Young and Duke's reports, there were more of them than anyone had expected. The term "plains" conjures up images of Kansan flatness. But at one point John Young said sarcastically, "Cayley *Plains?* There's nothin' 'plains' about this place." Instead, "there's just craters on top of craters." The craters made the driving rough, but even worse, Young couldn't see them as he headed west. With the sun at his back the landscape was as bright and featureless as a snowfield; all shadows were hidden by the objects that cast them. Muehlberger could hear him telling Duke, "This driving down-sun is murder." What worried Muehlberger—and Young—was that Elston and Boudette had mapped 10- to 16-foot-high scarps. So far, though, Young and Duke hadn't reported any.

Now Young and Duke were nearing their objective, a worn, 1,000-foot crater called Flag, that was big enough to have penetrated deep into bedrock. But how to sample the rocks it brought up? By chance, there was a much smaller (120-foot), more distinct crater, called Plum, right on Flag's rim; *its* rim was hoped to be littered with pieces of Flag's ejecta. Those rocks would show what the Cayley Plains were made of.

3:44 P.M.
PLUM CRATER

At Plum, Young and Duke picked up one rock after another and described them all as breccias. Muehlberger's team, wondering if they'd misheard, sent a message to Tony England to find out if perhaps they'd found a volcanic rock after all. England didn't want to nag Young and Duke; he was sure that if they'd found one, they would have said something. Meanwhile, Young approached a boulder, hammer in hand, and pried off a piece. The rock was somewhat weak and friable, not hard, like basalt; everyone listening knew that such weakness is typical of rocks that have been through an explosion from a meteorite impact. England asked, "Do you think it's a breccia?"

"Yeah, it's a breccia, Houston," said Young. "Or a welded—" Muehlberger

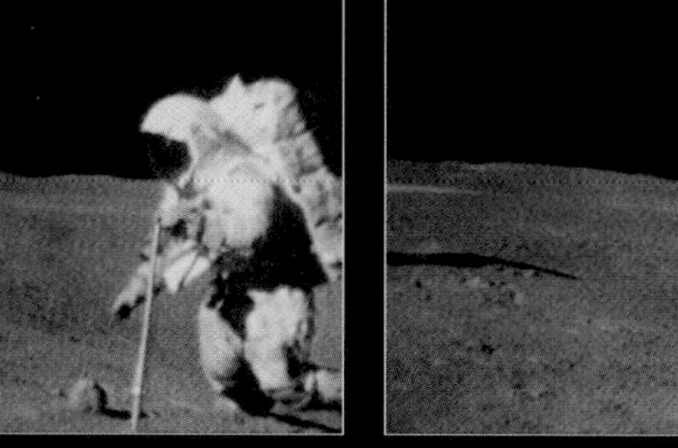

As Young and Duke return to the Rover at Plum crater, geologists on earth ask them to pick up a rock they've spotted on TV. Reaching the rock, Duke kneels down and rolls it up his leg. At nearly 26 pounds it would prove to be the largest lunar sample returned to earth by any Apollo mission.

could hear the words forming on Young's lips: a welded ashflow tuff, which is a rock made of ash fragments welded together by a matrix of volcanic glass. Welded ashflow tuffs can look just like breccias, and they are rich in silica, just as the Cayley rocks were thought to be. Some of Muehlberger's team had predicted there would be welded ashflow tuffs at Descartes.

"No, that's not right," Young said. "It's a breccia."

In the back room, the question was getting harder to ignore: Where were the volcanic rocks?

4:01 P.M.

Ever since Young and Duke arrived at Plum, Muehlberger's team had eyed a rock sitting on the crater's east rim, near the Rover and its TV camera. They thought they could see a crystal shining through a layer of dust, and that meant it might be igneous. They forwarded the request to mission control.

"Are you sure you want a rock that big, Houston?" asked Young, mindful of the 200-pound limit on the samples *Orion* could carry. "That's twenty pounds of rock right there." It turned out closer to 26 pounds, and it was bigger than a football; about a quarter of it had been buried in the dust. To get it, Duke had to get down on his knees, roll the rock up the side of his leg, and then, clutching it to his suit, try to stand up without losing his balance. "If I fall into Plum crater getting this rock," Duke said with mock annoyance, "Muehlberger has *had* it." It would prove to be the biggest moon rock ever brought back to earth, and it would be christened "Big Muley," in honor of Bill Muehlberger. The rock was so covered with dust that the men couldn't tell what it was, but Duke ventured that it might be a breccia.

4:47 P.M.
MISSION OPERATIONS
CONTROL ROOM

In mission control, Tony England listened to the communications loop as Young and Duke headed back toward *Orion* to end their first moonwalk. Charlie Duke had been using more cooling water than expected, and flight director Pete Frank had decided to trim the second and last geology stop of the day to only 19 minutes. Frustrated by the cutback, Young and Duke were thankful that the driving was easier as they headed east, because they could follow their tracks. At Buster crater, while Young took readings with a portable magnetometer, to record any remnant magnetism in the rocks of the

Filmed by the 16-mm movie camera mounted near the front of the Rover, Young adjusts the vehicle's television camera. The bag attached to his life-support backpack contains samples he and Duke have picked up during the moonwalk.

Cayley Plains, Duke had enough time to gather several more samples. Every one was a breccia. And as he listened, England could not help but feel apprehension for the men at Descartes.

One month shy of his thirtieth birthday, with a clear, serious face and wire-rimmed glasses, England looked as though he might be a junior professor in a college science department. In fact he had a Ph.D. in geophysics, the only geoscientist in the astronaut corps besides Jack Schmitt. Unlike many others in the XS-11, who had immersed themselves in Skylab activities in hopes of improving their chances of flying, England had chosen to stay with Apollo. He knew that in doing so he was probably giving up his chance to fly anytime in the near future, but his greatest fascination lay with the moon and the planets. If NASA had not canceled the last three lunar landings, England might now be looking forward to going to the moon himself. Instead, as Apollo 16's mission scientist, this seat in mission control was as close as he would get.

There were times when the earth-moon gap seemed especially vast to England, and one of them had been when the heat-flow cable broke. England had not been looking at the screen until it was too late to warn Young. Young hadn't said a word about it since the mishap, but he didn't have to; England knew he must feel absolutely terrible. So must Duke; he had put in a great deal of time training with the drill, and he'd really been looking forward to doing a good job for Mark Langseth. Now, despite the exuberance they'd shown at Plum, Young and Duke were a subdued pair as they drove past the ALSEP. England wanted to say something to cheer them both up. "A day ago, it didn't

In another 16-mm movie frame, Young brushes lunar dust from Duke's Hasselblad camera. The fine, abrasive particles were a constant nuisance to the men on the moon.

look like we were even gonna land," he radioed, "and now we've sampled our first highlands. I feel pretty good about the science without the heat flow."

"Well, I know Mark's disappointed," Duke said, "and I sure am." His voice was flat with disappointment.

"Me too," Young said.

England worried about Young and Duke's morale even aside from the heat flow accident. The rocks were turning out to be so different from what they had all expected. Were Young and Duke doubting themselves? Were they telling themselves that if they didn't find volcanic rocks, they weren't doing a good job? England could only imagine how big the silence at the other end of the earth-moon communications link must feel; he tried to fill it with reassurance. When one of the astronauts made a discovery, England radioed his own enthusiasm—*"Outstanding!"* And when one of them made a joke, he keyed his mike to broadcast his own laughter. He tried to communicate what he could not say over the air: "Don't worry. You're seeing something we didn't expect. Don't try to make it fit. See what you see and document that; we're going to have to put together a whole new picture of this place because of what you do see."

But England need not have worried about Young and Duke. They had seen enough breccias on field trips, and in the Lunar Receiving Laboratory, to recognize them without hesitation on the moon. They had also seen more volcanic rocks than most field geologists, and there definitely weren't any in sight on the Cayley Plains. And they never doubted themselves. In fact, it occurred to Charlie Duke that the geologists might be doubting what they were hearing. He could imagine them saying to themselves, "My God, we wasted three years of training on them." He hoped that wasn't the case.

DINING IN SPACE

Young and his crewmates, like their predecessors, were sustained by a variety of specially prepared foods. Freeze-dried meals, which made their first appearance on Gemini missions, were augmented in Apollo by wet-pack dishes such as beef and gravy *(below)*, warmed by being dipped in hot water. For Apollo 16, flip-top cans were introduced, including the serving of peaches sampled by Duke *(above)*. Potassium-spiked orange juice turned out to be Apollo 16's most problematic galley item. To prevent heartbeat irregularities, crews had to drink so much of the beverage that it became a source of gastric distress.

It wasn't. There were some perplexed scientists in the back room, and even a wounded ego or two, but if Muehlberger and his team were surprised by the descriptions from the moon, they did not doubt their veracity. Like Scott and Irwin before them, Young and Duke had won the geologists' confidence before they left earth. The simple fact that they were at the Descartes highlands, "field checking" the volcanic hypothesis, was exciting enough to overshadow the confusion of confronting the unexpected. Soon enough, they would have pieces of the Cayley Plains to study for themselves. On their monitors, Muehlberger could see Young, standing in *Orion*'s shadow, using the conveyor line to haul the day's rock box up to Duke. Would those samples disprove the volcanic hypothesis? Today's prospecting had turned up a few rocks that did not seem to be breccias, but they weren't clearly volcanic either. Maybe some welded ashflow tuffs would turn up after all. Or maybe the highlands were indeed covered by a surface layer of impact rocks—with so many craters around, that would not be surprising—and there were volcanic flows underneath. If so, the samples in that rock box, now in Duke's hands, would not give the answer. That would have to come from the mission's prime objective, huge North Ray crater, which the geologists hoped would be big enough to have blasted deep into bedrock. North Ray was shaping up as the single destination for a shortened third moonwalk. And tomorrow's excursion, to the flanks of Stone Mountain, would offer the first chance to probe the riddle of the Descartes mountains. That moonwalk was now threatened by the possibility that the astronauts might be called on to repair the broken heat-flow experiment; Muehlberger hated to think of the time that might eat up. In any case, Young and Duke would follow the traverse plan just as they had trained. True, those traverses had been meticulously designed to investigate a volcanic terrain, but there wasn't any way to reframe them now, and even if there had been, no one could think of any better plan.

8:44 P.M.

Meanwhile, back in *Orion,* John Young was worried about orange juice. The NASA doctors had been quite alarmed by Scott and Irwin's heart irregularities on Apollo 15, which they attributed largely to a loss of potassium. To prevent a recurrence on Apollo 16, they added electrolyte to the food, mostly in the form of potassium-spiked orange juice, and they instructed Young, Mattingly, and Duke to drink as much of it as possible. Already that had caused a problem for Charlie Duke. Back in lunar orbit, preparing for the landing, Duke had discovered that every time he moved his head within his bubble helmet, the microphone on his communications hat triggered the valve on the drink bag just below his neck. Before long, Duke found himself staring at a big blob of orange juice floating in front of his face, just out of reach. When it touched the microphone boom it attached itself and migrated under the fabric of his hat, into his hair. By the time he could take off his helmet, after landing, he looked as if he'd been shampooing with orange juice.

But the steady diet of potassium-spiked juice had even more unpleasant effects. A few minutes ago, Tony England had passed on word from the medics that the EKGs looked great, and added, "Just push on the orange juice."

Young drawled, "Push on the orange juice and everything will be fine?"

"Yes, push on the orange juice. Roger."

"I'm gonna turn into a citrus product is what I'm gonna do."

England suppressed a laugh. "Oh, well, it's good for you, John."

But Young wasn't kidding. "Ever hear of acid stomach, Tony? I think I've got a pH factor of about three right now. Because of the orange juice." Young was fed up with the havoc being wreaked on his digestive tract. Up in the command module the three of them had all but overpowered the environmental control system with methane. So far they'd kept their complaints to themselves, but now, as Young and Duke finished their housekeeping chores for the night, they didn't realize that Young's mike button was stuck in the "on" position. What came down from the moon was a backstage look at the John and Charlie show.

"I got the farts again, Charlie. I don't know what gives 'em to me, I really don't. I think it's acid in the stomach, I really do."

IN 1972

Three Japanese terrorists fire on a crowd at Lod airport in Tel Aviv, killing 28 and wounding 76.

Ceylon becomes the Republic of Sri Lanka under a new constitution.

Photojournalism pioneer *Life* magazine suspends publication after 36 years of weekly issues.

Only 22 of the 38 states needed to amend the U.S. Constitution ratify the proposed 27th Amendment.

"Prob'ly is," Duke said.

"I mean, I haven't eaten this much citrus fruit in twenty years. But I'll tell you one thing, in another twelve fuckin' days, I ain't never eatin' any more. And if they offer to serve me potassium with my breakfast, I'm gonna *throw up.* I like an occasional orange—really do. But I'll be damned if I'm gonna be buried in oranges."

England beeped his mike button several times to signal the performers, but they kept talking—no longer about oranges, but about the moonwalk they'd just finished. Finally, England spoke up: "*Orion,* Houston."

A crisp response from Young, still unaware: "Yes, sir!"

"Okay, John, [you] have a hot-mike."

Young, more subdued now: "How long—How long have we had that?" No one within the sound of Young's voice would forget the episode, including Deke Slayton. He had been sitting next to Tony England throughout the day, keeping watch on the mission, as he had with every important moment of every flight since Young and Gus Grissom flew the first Gemini mission. Now, as Young and Duke prepared for sleep, Slayton manned the Capcom mike for a short time. With his lined face and western shirt he looked like a cowboy who had seen many hard winters. When he spoke to *Orion,* the men on the moon were glad to hear his voice, and a brief chat ensued. Duke talked about the previous night, when he had lain awake, overcome by excitement and anticipation, while Young sawed wood. "Couldn't believe we'd go to sleep, Deke, but man, this guy John sleeps like a baby up here."

"It sounds like the best place in the world to sleep," Slayton said simply. "I wish I was with you."

Young answered, "We do too, boss."

II: "YOU JUST BIT OFF MORE THAN YOU CAN CHEW"

SATURDAY, APRIL 22
MANNED SPACECRAFT CENTER

It had been a long night for Mark Langseth and his heat-flow experiment. Shortly after the first moonwalk was finished, a task force had convened in a back room, including engineers from the Bendix Corporation, who built the experiment; Fred Haise, Apollo 16's backup commander; and a couple of people from the "Tiger Team," the squad of ever-ready engineers who were called into service to solve the knottiest hardware problems. Langseth was amazed by them; they soaked up information; they were inexhaustible; they were so confident that Langseth felt they could have surmounted any failure. The Tiger Team brought a five-gallon jug of ice cream and went to work. The cause of the accident became clear once the Bendix people described the connector that had attached the heat-flow cable to the Central Station. Originally it had been hard-wired; but at a late date Bendix changed the design to a connector. Langseth never saw the connector they used, but now they told him that it had a sharp edge; the cable had probably broken because the connector had sheared it in two. Langseth wasn't angry at John Young; he was furious at Bendix.

But how to recover? The cable could be reinserted into the connector, but first the astronauts must have a way to strip off its insulation to expose the wires. What was needed was an abrasive of some kind, and it just so happened that a very good one was available: lunar rock. By morning (and many scoops of ice cream later) the team had come up with a plan. Young and Duke would bring the cable and the still-attached heat flow electronics box

The Tiger Team brought a five-gallon jug of ice cream and went to work.

into the lunar module at the end of the second moonwalk. Inside the lander, they would wrap the end of the cable around the handle of the geology hammer and scrape off the insulation with a rock. Finally, they would cut a clean edge with the scissors. On the third moonwalk, they would reconnect the experiment to the Central Station. Fred Haise tested the procedure; it worked.

Around nine o'clock on Saturday morning, Langseth and the team went to the VIP room at the back of mission control to make their case to Chris Kraft and Rocco Petrone. A few other ALSEP scientists went along for moral support. No one needed to say that the astronauts would go along with the plan; fixing things was part of their being. And Langseth pointed out that a repair would demonstrate the value of having humans in space instead of machines (a point he hoped would play better than "I want my experiment fixed"). Kraft and Petrone listened, and they allowed as how it might work. But it was hardly a sure thing; there were forty-eight separate wires in that cable that had to make contact, and no one could guarantee the procedure would work on the moon. And even if it did work, it would cost a lot of time; the team estimated a total of an hour during the two moonwalks, plus another hour for the work inside *Orion*. Petrone said, "We're not going to do it." Langseth would say later that he did not question the decision. The astronauts' time was too precious to spend on fixing one experiment; they had the moon to contend with.

12:03 P.M., HOUSTON TIME
6 DAYS, 0 HOURS, 9 MINUTES MISSION ELAPSED TIME
DESCARTES HIGHLANDS

In one way, Stone Mountain was just like Hadley Delta: it was steeper than either Young or Duke had realized on the way up. Young parked the Rover in a small, subdued crater so that it wouldn't start sliding down the hill the way Scott and Irwin's had. When he turned to look back to the north, he saw the mountain drop away from him in a series of steep-sided ridges. Off to the west it looked really treacherous, even more steep than the slope they had driven up. Young said to himself, "You've just nearly bit off more than you can chew." But the view from this high vantage—like the one that had confronted Scott and Irwin—was dazzling. They were 500 feet above the valley floor, higher than any Apollo moonwalkers ever had been or would be again. And the most spectacular sight was South Ray crater, five times the size of a football field, its rim an absolutely *brilliant* white, seemingly as fresh as the day it was blasted out of the Cayley Plains. Rays of boulders—black boulders, white boulders—were sprayed across the landscape for miles in all directions. The boulders next to the rim looked to be 90 feet across. South Ray was a beautiful excavation into the Cayley, and Young had pushed hard to go there on one of the moonwalks—even to add a fourth excursion for the purpose, if necessary. But astronomers had made soundings with radar beams

Five hundred feet up the side of Stone Mountain, Young works at the Rover. Around him are rocks that he and Duke suspected had been blasted from the valley floor by the impact that formed South Ray crater.

from earth, and when they analyzed the echoes from South Ray crater they determined there would be too many rocks for a Rover to gain safe passage. Now Young could see that they had overestimated the hazard. Still, on the way up here, he and Duke had driven across a rise with so many boulders that he feared one of them would break a wheel assembly off the Rover. Perhaps getting to South Ray would have been tough, but he felt sure they could have made it to its nearby, smaller counterpart Baby Ray. Under his excitement, Young felt the frustration of discoveries beyond his reach.

But if they could not visit South Ray, its rocky artillery fire was scattered across the southwestern face of Stone Mountain. Those rocks—clean and sharp-edged—set Stone apart from the bland slopes of Hadley Delta. And if the rocks really were South Ray ejecta, as Young suspected, then even though they were 500 feet up on Stone Mountain, they were picking up pieces of the

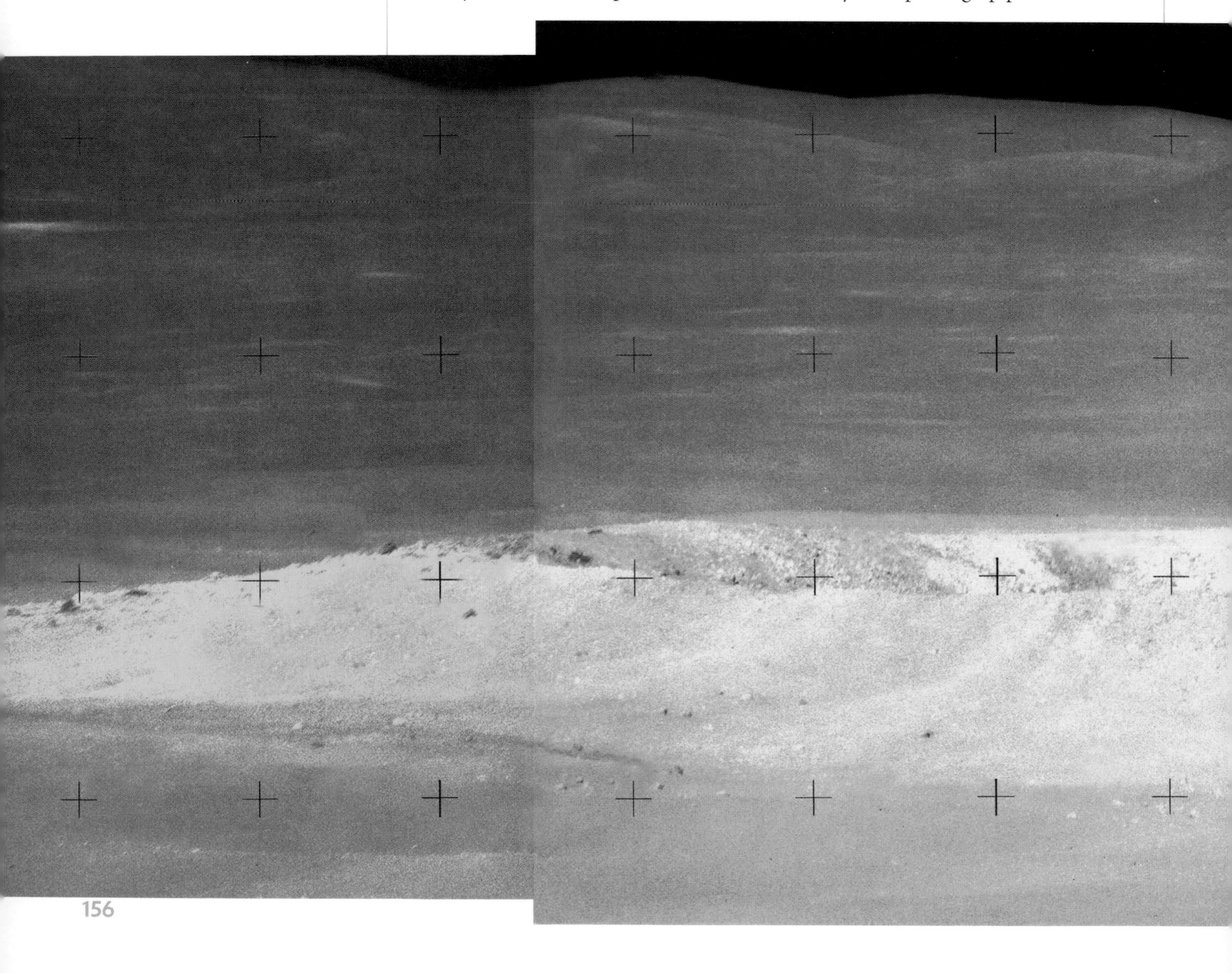

Cayley Plains. "You know, John," said Duke, "with all these rocks here, I'm not sure we're getting Descartes."

"That's right; I'm not either." When these words reached the back room, they only confirmed what Muehlberger and his team had anticipated before the mission. Realizing that Young and Duke might find South Ray ejecta here, they'd chosen five craters, deep enough to have penetrated bedrock, for Young and Duke to visit; they were named the Cincos. If Young and Duke could get to one of the Cincos, they would probably find large boulders whose source could be confidently tied to the mountain. Young and Duke had been looking for the Cincos on the drive up, but they were never sure where they were. Ironically, the largest of the five, Cinco *a,* was only 40 yards from the place where they stopped. Young and Duke would never realize that it was just out of view behind a ridge.

Using a telephoto lens, Duke photographed South Ray crater from Stone Mountain. Mission planners considered it unwise for Young and Duke to visit the crater, 1,500 feet in diameter, because boulders surrounding it posed too great a hazard to the Rover.

There weren't any large boulders here, but Young wasn't giving up. "Okay, Houston. I'm digging an exploratory trench." According to the geologists' theory, the mountain could have formed from sticky lavas that oozed to the surface and congealed, or cinders that flowed across the landscape and then solidified. The rocks Young plucked from the trench were covered with dust, but as far as he could tell they looked just like what he had seen down on the Cayley Plains. "I wish I could say these rocks look different, Houston, but they don't."

Like a prospector panning for gold, Young stood on the crater's soft, powdery wall and hacked away with the rake, struggling to hold the tool with tired, aching hands.

Now Tony England sent word from the back room to look for another crater within walking distance that might yield samples of Descartes. Young made his way about 50 yards up the slope to a 60-foot-wide, block-strewn bowl. When he reached the rim, he realized these rocks too were invaders from South Ray. Young had another idea that showed his insight: He decided to rake the side of the crater facing *away* from South Ray, the side that would have been shadowed from the barrage of ejecta. Like a prospector panning for gold, Young stood on the crater's soft, powdery wall and hacked away with the rake, struggling to hold the tool with tired, aching hands. His difficult harvest was not what he expected; what looked like rocks turned out to be clods of dirt that fell apart in the sample bag.

Years later, Young would say that when you are on the moon, inside a pressurized space suit, with only six or seven hours outside, and no more than an hour at any given place, you just can't take time to try and see the big picture. You're supposed to be getting the samples and documenting them, according to the timeline. If you try to be anything more than a technician, you are doing a disservice to the scientists who sent you. There was no anxiety on Stone Mountain now, as Young and Duke came bounding down the hillside, laughing as they returned to the Rover. The tension was in the back room, where Muehlberger and his team knew they had one chance left to snare samples of the Descartes on Stone Mountain.

After a nervous drive downslope, Young parked next to a 50-foot crater that seemed to be free of South Ray ejecta. Young again suggested that he and Duke work the slope that faced away from South Ray. After some effort—the men felt as though they were on the verge of tumbling into the crater—they culled from the crater wall some small whitish rocks that were not jagged like South Ray ejecta but rounded; to Young that meant they were older, having been eroded by micrometeorite rain. But even these were not clearly pieces of Descartes. He told Tony England, "I don't think this is going to be a simple problem, even after you get the rocks back. . . ."

But when the men climbed out of the crater and headed back to the Rover, they got the biggest surprise of the day; Young almost tripped over it. A shoe-sized white rock sparkled in the sunlight. "We're gonna get that one," said Young excitedly. "That's the first one I've seen here that I really believe is a crystalline rock." It seemed to be made entirely of plagioclase, much like the Genesis Rock, but with crystals that were tiny, like sugar. Like the Genesis Rock, it could be a surviving fragment of anorthosite from the primordial crust, but it had clearly been kicked around by eons of meteorite impacts. But it surely was not volcanic. And as Young and Duke headed down from Stone Mountain it seemed that on this mission there would be no moments of clear discovery like the one that had heralded the Genesis Rock, no excited announcements from the moon—*We found what we came for!* Instead, Young and Duke had found what was there. And in the back room, Muehlberger and his team felt it had been a good day. Young's ingenuity had greatly impressed them. Whatever the Descartes was made of, they knew Young and Duke had probably succeeded in collecting representative samples. The true nature of the Cayley and the Descartes was unfolding word by word, rock by rock.

III: ". . . OR WHEREVER GEOLOGISTS GO"

No one who knew Ken Mattingly would accuse him of being a patient man, but in the last two years he had sorely needed patience, and even more, raw persistence. After getting bumped from Apollo 13 a week before launch, Mattingly dove into two more years of training, with no letup, for Apollo 16. Then, with just two weeks to go, the NASA doctors told him they'd found an

Watching the Rover's television pictures of the astronauts, earthbound geologists see Young, taking panoramic photographs from Stone Mountain *(top)*, joined by Duke to collect samples from the side of the mountain *(middle)*. In the third frame, the pair carry their haul downslope to the Rover.

irregular reading in one of his blood tests; he had an elevated level of bilirubin, an indicator of liver function. Physically, he felt fine; in fact, he was in superb condition. But as far as he could tell, the doctors were suggesting he might be coming down with hepatitis. Surely they understood that this was an impending *disaster*—and yet they offered him no advice on how to avert it. They just kept taking blood tests, waiting to see what would happen. For three or four miserable days, Mattingly feared another medical false alarm was going to steal his last chance to go to the moon. Then the doctors decided he was fine.

Needless to say, it had all been worth it. The past seven days had been the climax of his career, and even aside from the flying, he had been living through one unforgettable sight after another. It got so he didn't want to look at each new spectacular for fear of erasing the memories of what had come before. Then, after the trauma of the wave-off, he was finally ready to carry out his solo mission. He knew the moon so well that he didn't even need to look at a map. It had been Mattingly who had been first to sign on with Farouk El-Baz. But even now, after working with the man for almost three years—and he liked El-Baz a great deal—he could not honestly say that he was interested in geology, or in the moon. And he had told El-Baz that, up-front. But he was ready to do whatever El-Baz asked, not just because he admired him, but because he wanted to be something more than a truck driver. He wanted to make a contribution—hell, he wanted something to do.

But what was he going to contribute? Would he really be able to see anything with his own eyes that wasn't already in the thousands of Lunar Orbiter pictures, or the Apollo photos? Oh, Farouk told him that there was no substitute for the human observer, but he wasn't unbiased, was he? After all, if there were no need for the human observer, then NASA wouldn't need Farouk!

But now, with three days in lunar orbit under his belt, he had to admit he could see more than he'd ever expected. What had looked like a pile of whitish rubble was now brimming with details. He found himself staying awake well past the start of his sleep periods because he didn't want to miss anything. Whatever the Cayley formation was made of, there was more of it all over the far side of the moon. To Mattingly, the bright, smooth highland plains didn't look volcanic; they seemed to consist of debris that had been shaken, like a bowl full of gravel, until the surface was relatively flat. He would tell the geologists all of this when he was back home. Still, Mattingly knew, you can't analyze the dimensions, or the brightness, or the color, of an impression. The real data would be the pictures: cameras don't make mistakes; he couldn't say the same for himself.

SUNDAY, APRIL 23
10:39 A.M., HOUSTON TIME
6 DAYS, 22 HOURS,
45 MINUTES MISSION
ELAPSED TIME

Inside *Casper,* in the pitch blackness of an orbital night, Ken Mattingly heard his own voice, on tape, break the stillness. "In twenty seconds the DAC will go on and remain on until sunrise. Adjust the settings." Outside, the depths of space were crossed by fingers of light, streamers of the sun's outer atmosphere. It is this envelope of dimly glowing gas, the corona, that frames the moon's silhouette during a total solar eclipse on earth. For a glimpse of its cold, eerie light astronomers will travel halfway around the world, but Mattingly now saw the corona as only the space traveler could, in the last minutes of orbital night, while the sun still hid below the unseen horizon. It was Mattingly's task to capture the corona on film using the Data Acquisition Camera (DAC). And the tape—that was a matter of efficiency. Mattingly knew the inside of the command module so well that even in pitch darkness he could find his way around. And he knew that if he flicked on a flashlight to glance at a checklist, even for a moment, he'd ruin his night vision. So he'd spent an hour during the trip out to the moon reading his checklist into the portable tape recorder.

"Stand by for the start. Four, three, two, one, *start.*" Mattingly activated the camera and heard the steady click of advancing film. Data: that's what it was all about. Slowly the luminous coronal streamers brightened, and then pitch darkness yielded to blinding sunlight. And now came the best time of all. Even though he would have traded places with Young and Duke in an instant, he could not imagine how bounding across the surface of the moon was more fun than orbiting it alone. Charlie Duke liked country music, and he'd brought along a tape of "Grand Ole Opry" stars from the same guy who'd supplied Pete Conrad and Stu Roosa. He played it all the way out to the moon. Mattingly almost got to like country—almost. But now, with *Casper* to himself, Mattingly savored the music he loved. In Houston they

In the geology back room, Bill Muehlberger *(seated at center)* prepares instructions for relay to Young and Duke. At his left are Jim Head *(standing)* and Dale Jackson *(seated)*; to his right, standing by the doorway, are Apollo 17 prime crewmen Jack Schmitt and Gene Cernan.

could hear snatches of it in the background whenever he keyed his mike: a Mahler symphony, Holst's *The Planets,* and Berlioz's *Symphonie Fantastique.* ("It didn't sound as good as 'Ridin' Old Paint,' " teased Stu Roosa, "but I guess it'll do.") And here, over the far side, it was the perfect background music to the silent, unreal panorama that filled the windows. He went about his work in a carefully orchestrated, weightless ballet, tending the scientific instruments, taking pictures, and turning his eyes moonward in search of a new discovery for Farouk El-Baz.

"And there's old Mother Earth," Mattingly said aloud. "Man, that's a beauty, too. Never get tired of watching earthrise." Earth: the source of all his troubles. To no one's surprise Mattingly had put together a staggering solo flight plan and was bent on doing it all. He wouldn't waste time checking the mission clock for the timing of his tasks. Instead, whenever he was in radio contact, he would let the Capcom watch the systems and the clock for him, leaving him free to concentrate on the moon. When Houston needed him to turn an experiment on or change a switch setting, he was a happy robot. And if they'd only left him alone, he would have been glad to continue that way for another two days. But the engine problem, and then Young and Duke's late landing, seemed to have thrown all of mission control into a tailspin. For some reason—he still wasn't sure why—they were going to cut an entire day off the mission, one of the two days the three men had planned to spend studying the moon after the surface work was finished. That alone wreaked havoc on Mattingly's flight plan. He'd specifically talked to the flight directors before the mission, and they'd agreed that if something unexpected should interfere with the schedule, they wouldn't try to reschedule everything; they'd just lop off the parts that were affected. But the flight directors seemed to have forgotten all about that. Every time he turned around there was another revision to copy down. Even more frustrating, the flight planners were sending him procedures that hadn't been tested; they hadn't had time to check their own work. It was as if they'd never flown a mission before. They were ruining his ballet. He wanted to get on the radio and say, "Alright, you guys, knock it off." Instead he called up every ounce of self-restraint and kept his irritation to himself.

Even when he was out of their reach, over the far side, he fell prey to the same problem that had plagued his predecessors: too much to do. Before the flight, without even realizing it, he'd let extra tasks creep into the time reserved for eating, and he didn't realize what that would be like until now. As it was, just about the only time he managed to eat was when he had to stop to go to the bathroom. There he was, slurping down a plastic bag of juice while hooked up to the urine collection hose, with a fecal collection bag flypapered to his rear end. Not the image of the Intrepid Lunar Explorer. Mattingly always wanted to be the first man to go to Mars. If that trip was going to be anything like Apollo, he'd tell them to forget it.

SUNDAY, APRIL 23
ON THE CAYLEY PLAINS

Like most astronauts, Charlie Duke would tell you that he rarely dreams at night, and remembers his dreams even less often. But Duke had a dream, six months before the flight, that he had no trouble remembering. In Hawaii on a geology trip, Duke came down with the flu and ran a high fever; at times he was almost delirious. Some of the wives had come along, and Duke's wife, Dotty, took care of him while his colleagues were out in the field. In a fever sleep, Duke saw himself and John Young on the moon, driving their Rover toward North Ray crater. They came up over a ridge and suddenly Duke spotted something that made his heart race. A set of tracks crossed the ground ahead. Young stopped the Rover and they got off to investigate. The imprints in the dust looked like those from the Rover, but they were definitely different. Duke asked mission control, "Can we follow the tracks?"

"Go ahead," was the reply from earth. The twin trails stretched eastward, and Young and Duke turned to follow them. They drove onward for miles, over hills and across craters, until finally, topping another rise, they saw it: a vehicle, looking amazingly like the Rover, stopped on the surface. Aboard were two figures in space suits. After calling Houston to announce their incredible discovery, Young and Duke climbed off the Rover and approached

The crater walls plummeted steeply to a rocky floor 200 yards below, and the ground at the rim might be ready to give way at any moment.

the two figures, motionless in their seats. When Duke reached the one in the right seat, he could not see into its helmet because of its opaque sun visor. He put out his hand and raised the visor and saw his own face. The one in the left seat was John Young's double. After taking pieces of the space suits and Rover at mission control's request, Young and Duke drove back to the LM and blasted off for home. The next thing Duke knew, he was on earth, presenting the samples to the scientists. The test results: The craft was 100,000 years old. Then he woke up. The dream was so vivid—not scary, just *real*—that Duke remembered it from then on, and as he descended to the real moon inside *Orion,* he glanced out to his right at North Ray crater, and scanned the ground not only for boulders—"Looks like we're gonna make it, John; there's not too many blocks up there"—but for a set of tracks.

10:32 A.M.
SCIENCE OPERATIONS ROOM

Consider the athletic standing of the Caltech football team (low), and understand that Bill Muehlberger had seen his share of tough games. By any objective measure, Apollo 16 had turned out to be one of them; in the contest between the geologists and the moon, the moon clearly had the upper hand. Yesterday, Muehlberger had ruefully realized that the astronomers' radar soundings of South Ray crater—so promising for samples from deep within the Cayley—had misled them all into thinking it was inaccessible. And today the clock was running out on their chances to find volcanic rocks in the Descartes highlands. The managers had trimmed Apollo 16's final moonwalk to only 5 hours—no extensions, lest they violate the liftoff time—leaving just enough time for a dash to North Ray crater, some 3 miles north of the lunar module. North Ray crater was a prime destination for many reasons. For one thing, no astronauts had yet reached the rim of a large, well-preserved crater to hunt for samples of deep bedrock. North Ray certainly fit the bill; not only was it fresh, but it was more than six-tenths of a mile across—nearly as big as Arizona's Meteor crater—and more than 650 feet deep. North Ray was big enough to have punched through any surface layer of soil and debris into the underlying Descartes rocks at the base of Smoky Mountain. The explosive impact that formed it had sprinkled boulders across the landscape like grains of sand. The ones on North Ray's rim might have come from hundreds of feet down, and some of them were so big that they showed up on the photomaps. Young and Duke had already spotted a few of those boulders during the drive up. As he waited for them to reach the crater, Muehlberger was not grim. If his team had known a little confusion in the past two days, then their two surrogates on the moon had more than made up for that with their skill and insight. Muehlberger was proud of them.

10:37 A.M.

"Oh, spectacular! Just spectacular!" Young's excited voice filled the back room; he and Duke had reached the crater. Soon a TV picture appeared on the monitors, but Young and Duke weren't in it. Ed Fendell, the engineer who operated the Rover's camera from his seat in mission control, had aimed it at a blue-and-white crescent in a black sky. "C'mon, camera," Muehlberger said impatiently, "quit looking at the earth. Goddammit." At last the camera panned down to the sight of two tiny space-suited figures standing before a gigantic pit, which was so big that it extended well beyond the field of view. Everywhere, rocks poked through a mantle of dust. To some of those watching, including Tony England, this was the most nerve-racking moment of the mission. The crater walls plummeted steeply to a rocky floor 200 yards below, and the ground at the rim might be ready to give way at any moment. If one of the astronauts fell in, he would never get out. But the picture was deceptive, just as it had been with Scott and Irwin at Hadley Rille; the men were far from the edge. The geologists had hoped Young and Duke would be able

to get close enough to the crater to see the bottom and look forexposed bedrock. Now they realized that was out of the question.

"That rascally rim—it slopes [toward the edge] about ten or fifteen degrees, which is the kind of slope I'm standing on right now," Young explained. "And then all of a sudden, in order to see to the bottom, I've got to walk another hundred yards down a twenty-five to thirty degree slope, and I don't think I'd better." Before the flight, Young had talked about bringing along a 100-foot tether so that one man could venture to the rim, or even

If Young and Duke couldn't see the bottom of North Ray, they could still bring back a piece of it.

part way into the crater, while the other stayed behind to anchor him. The tether never made it onto the stowage list, not only to save weight but because the idea made the managers too nervous. Now, without that line, Young wasn't about to let either of them get anywhere near the rim.

The men backed off and spent forty minutes taking pictures of the crater walls and collecting samples. While they worked, the TV camera panned the bright ground near the crater's eastern rim, and there, far in the distance, was a black shape. "Good Lord," said one of the geologists, "is that a boulder?" Indeed it was; Muehlberger could see it on his photomap, and it was dark, like the rocks on the crater floor. If Young and Duke couldn't see the bottom of North Ray, they could still bring back a piece of it—provided they got to the boulder in time. Little more than an hour was allotted for this stop; much of it was already gone.

Young and Duke had been eyeing the big boulder from the moment they arrived. Now, finally, Young said, "Okay, Charlie. Let's go back to the Rover. Put your bag on there and head out for the big rock." He cast his gaze toward the boulder and said with amazement, "Look at the *size* of that biggie. It may be further away than we think."

"No, it's not very far," promised Duke.

But by this time Young had had enough experience with trying to judge distance on the moon that his lunar module pilot's words didn't convince him. "Theoretically, huh? Like everything else around here: 'A couple of

At North Ray crater, Young uses a specially designed rake to sift cobble-sized moon rocks from the fine dust and smaller pebbles of the lunar surface. The Rover is parked in the background, next to a large boulder.

Televised views from the Rover show Young and Duke approaching a dark boulder near the rim of North Ray crater *(top)*. As the camera zooms in *(middle)*, the boulder's huge dimensions become apparent. Reaching their goal *(bottom)*, Young and Duke are dwarfed by the mass of stone, which they named House Rock.

weeks later . . .' " The pair began to lope away from the Rover, Duke leading the way.

On the monitors, Young and Duke ran onward, getting smaller and smaller. It began to dawn on the geologists that this boulder was even bigger than they thought. Suddenly, from the back of the room, Jack Schmitt's voice brought an eruption of laughter: "And as our crew sinks slowly in the west . . ."

Still Young and Duke ran. On the monitors they were tiny. The geologists heard amazement in Duke's voice: "Look at the size of that *rock!*"

"They're not even *there* yet," Muehlberger said quietly. Then, at last, Young and Duke stood next to a wall of dark, rough stone. The boulder was as tall as a four-story building and twice as long.

"Well, Tony," said Duke, "that's your House Rock, right there."

"House Rock?" asked a voice in the back room.

"House Rock," Muehlberger repeated quietly, like Ahab sighting the White Whale. If anything was going to show what the Descartes highlands were made of, it was House Rock.

"Okay," Young advised, "we had to come down a pretty good slope to get to this rock, so we may have to leave early to get back."

Schmitt looked at Muehlberger. "Get ready to cut other time somewhere," he said. "They're going to be here awhile." But there weren't going to be any extensions here. The walkback limit was immutable.

"You've got about seventeen minutes before you'll have to drive off," Tony England radioed, "so we'll have to hustle with this."

House Rock was dark, like basalt, but it didn't take Young and Duke long to see that it wasn't basalt; it was an enormous breccia, with fragments that were more than 6 feet across. Young and Duke weren't about to say so, but here was the last nail in the coffin for the volcanic hypothesis, hammered in by a rock the size of a house.

MONDAY, APRIL 24
11:13 P.M.
ABOARD *CASPER*,
HEADING BACK TO EARTH

Taken out of lunar orbit a day early—they were sure the managers were just too nervous to wait any longer to bring them home—Young, Mattingly, and Duke sped away from the moon. Like those who came before them, they recorded the unreal view with their TV camera, but the pictures went no further than mission control. Even during the moonwalks the networks didn't broadcast for more than a couple of hours at a time. Duke's parents had to go to mission control's VIP room to see their son at work on the moon.

At the Capcom's suggestion they turned the camera on themselves so that their wives could get a look at them on tape when they came in the next day. When it was Young's turn to be on camera, he held a smudged hand before the lens. "See that? Can you see the dirt under those fingernails? That's moon dust. You talk about two dirty people. It took us ten minutes before we could get Ken to open the door. And we're still that way." He wasn't exaggerating; even now, little gray pebbles drifted around the cabin, and packed away in the rock boxes were a whopping 207 pounds of samples. "Yeah," Young told his audience, "wait 'til you see some of those rocks."

But Muehlberger and his team couldn't wait. They were under some pressure from the media to explain the surprising findings. The nature of the Descartes was still an open question, but for the Cayley Plains, Muehlberger's team was throwing out the volcanic theory and resurrecting an older idea that went back to the cataclysmic blast that formed the Imbrium basin. The geologists envisioned fast-moving debris surging away from the newly formed basin and across the face of the moon. Reaching the Descartes highlands, more than 600 miles away, it filled the valleys between the mountains to form the Cayley Plains. When Tony England told Young and Duke about

Young and Duke stand at the rim of North Ray crater in this mosaic of views from the Rover's TV camera. With a diameter of 3,100 feet and a depth of 650 feet, the crater could hold tens of thousands of House Rocks *(opposite).*

His once-pristine space suit grimy with lunar dust, Young collects samples at North Ray crater. Nearby, the Rover displays its peculiar see-through tires, which support the vehicle by means of springy wire mesh instead of rubber filled with air.

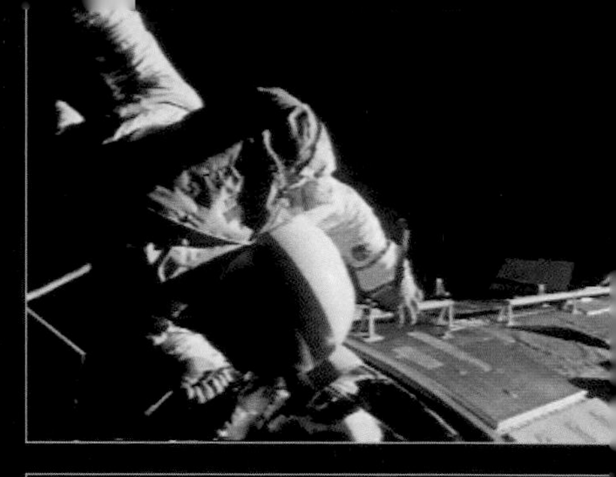

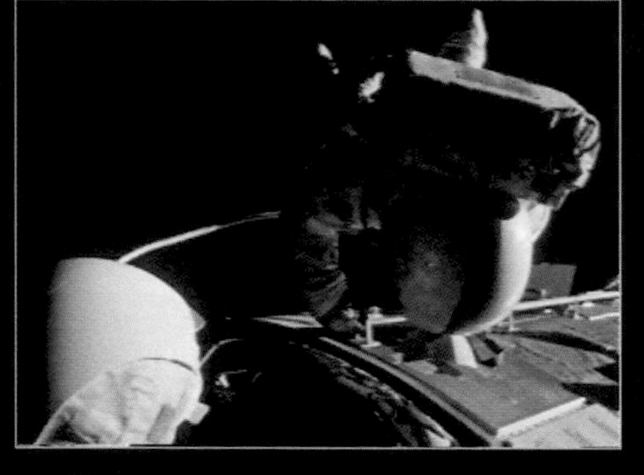
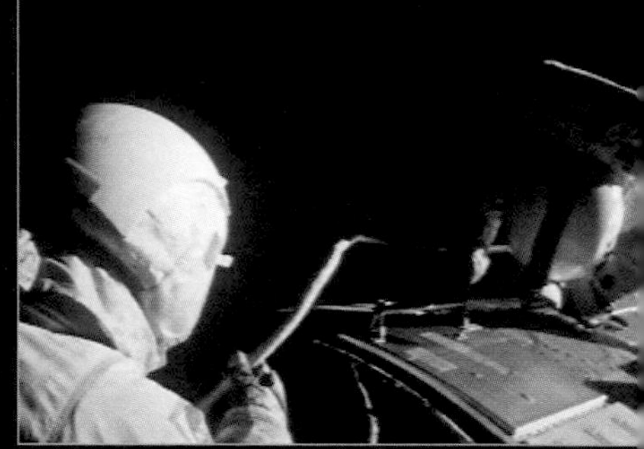
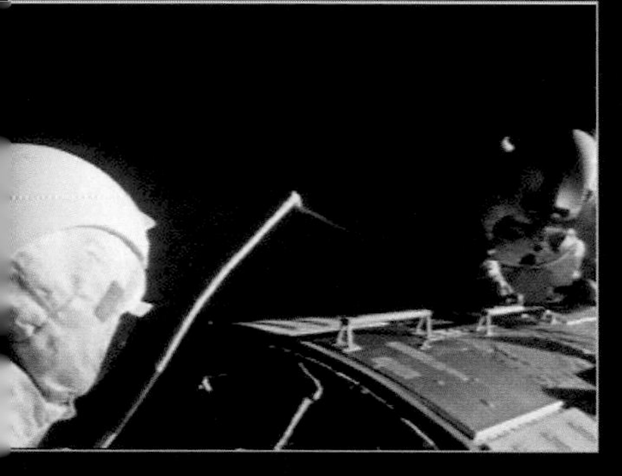
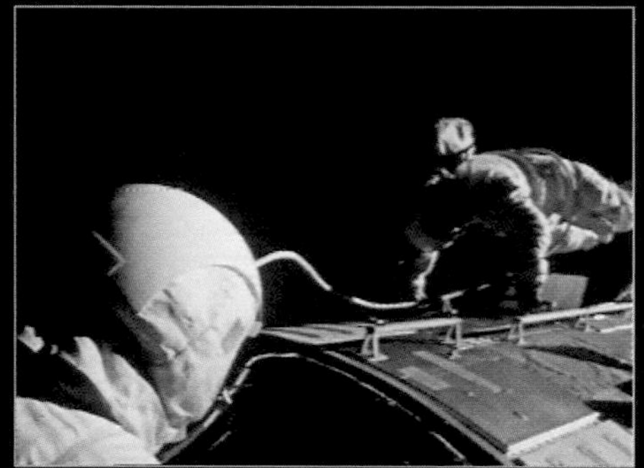
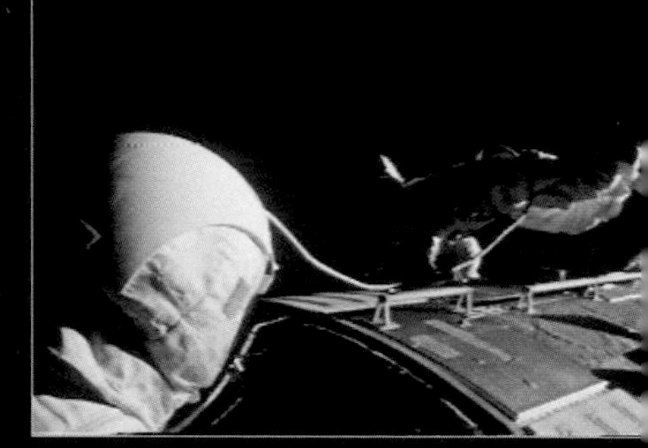

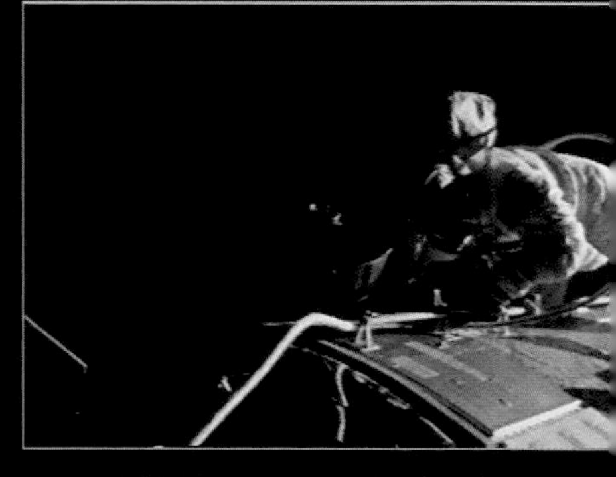

the theory of a "slosh" from Imbrium and asked whether they had any response, Young tried to be diplomatic. It was too early to be saying something like that, he said, too early to be jumping to conclusions about the geology when they hadn't even seen the rocks. He said simply, "It ain't good science." The geologists would simply have to be patient. The world's only experts on the Cayley and the Descartes were on their way home, and their credentials were under their fingernails.

TUESDAY, APRIL 25
2:43 P.M., HOUSTON TIME

For Ken Mattingly, the understanding of what it means to be in space did not begin the moment he reached earth orbit, or saw his world shrink to the size of his thumb; it did not come to him in solitude above the far side of the moon. It came, instead, during the trip back to earth, when he opened *Casper*'s side hatch, heart pounding with excitement, and floated outside. His mission was to retrieve two canisters of exposed film from the side of the service module. Training for the spacewalk he'd spent hours immersed in a water tank, clambering around in a vacuum chamber, and frantically practicing his tasks in the KC-135, a converted cargo plane that created about half a minute of weightlessness by flying in a parabolic arc. By the time Mattingly climbed into his suit aboard *Casper* and readied for the trip outside, he had

Homeward bound after three days circling the moon, Ken Mattingly takes a space walk *(left)*, recorded by a movie camera mounted on the command module's hatch. His mission: to retrieve film canisters from the service module's SIM bay *(page 86)*. In seven of the frames, one of which is enlarged at right, Charlie Duke is seen floating in the command module's side hatchway to assist his crewmate.

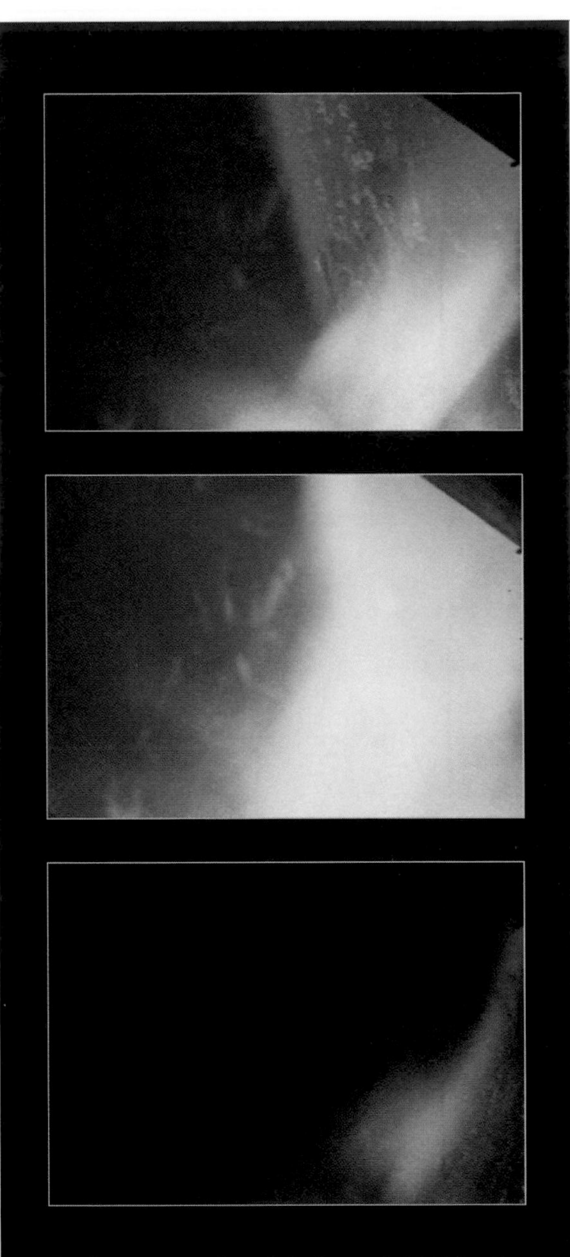

During command module *Casper*'s 25,000 mile-per-hour reentry into the earth's atmosphere, the window-mounted movie camera recorded a fiery glow from the spacecraft's heat shield as it protected *Casper* in the final stage of its flight *(top)*. As reentry progressed, the glow intensified *(middle)*, then subsided and changed color as the spacecraft slowed *(bottom)*.

anticipated every aspect, except the experience of being in deep space. ☾

The sun was so staggeringly bright that Mattingly immediately pulled down his gold-plated outer visor. He heard the reassuring whoosh of oxygen flowing into his suit through the 50-foot umbilical. He was completely outside now. On the silver and white skin of the cylindrical service module, he saw that here and there the paint was bubbled, from the heat of the maneuvering thrusters. The scientific instrument bay was near the other end of the cylinder, and he made his way there by "walking" with his hands along a handrail. It was effortless, just the way it had been in the zero g airplane. Except that it wasn't like the airplane, because everywhere he looked, beyond, around, and past the service module, there was *nothing* at all, and he realized that this machine he was holding onto—for nine days, his universe—was a speck in the void. He squeezed the handrail so tightly that if it hadn't been for the gloves, he would say later, he would surely have left fingerprints.

Arriving at the scientific instrument bay, Mattingly slid his boots into a pair of slipper-like footholds and rested. Looking back along the bright hull he saw Charlie Duke standing in *Casper*'s hatch, tending the umbilical. Beyond Duke, just off the nose, a full moon glowed, 50,000 miles away. When he looked to his left, he saw a tiny crescent earth, 180,000 miles away. Looking at them through *Casper*'s windows, he had never sensed the emptiness that lay on the other side of the glass. And in there, he had seen stars: *Where were all the stars?* It was a three-dimensional abyss. Charlie Duke kept saying, "My God, it's *dark* out here!"—and each time, Mattingly laughed, but his heart raced.

Mattingly was sure the "disappearance" of the stars was due to his gold visor. The doctors had advised him to leave the reflector down, lest he be exposed to harmful solar radiation, but he couldn't stand it anymore. He blinked the visor open just long enough for the universe to show a familiar face: *There they are!* His work finished, Mattingly pulled *Casper*'s hatch shut, and his universe became once more a small spacecraft drifting toward earth.

When Charlie Duke stepped off the ramp at Ellington, there was NASA geologist Fred Hörz coming to greet him out of the crowd of well-wishers. Duke grinned, "Those were sure funny-looking volcanic rocks, Fred."

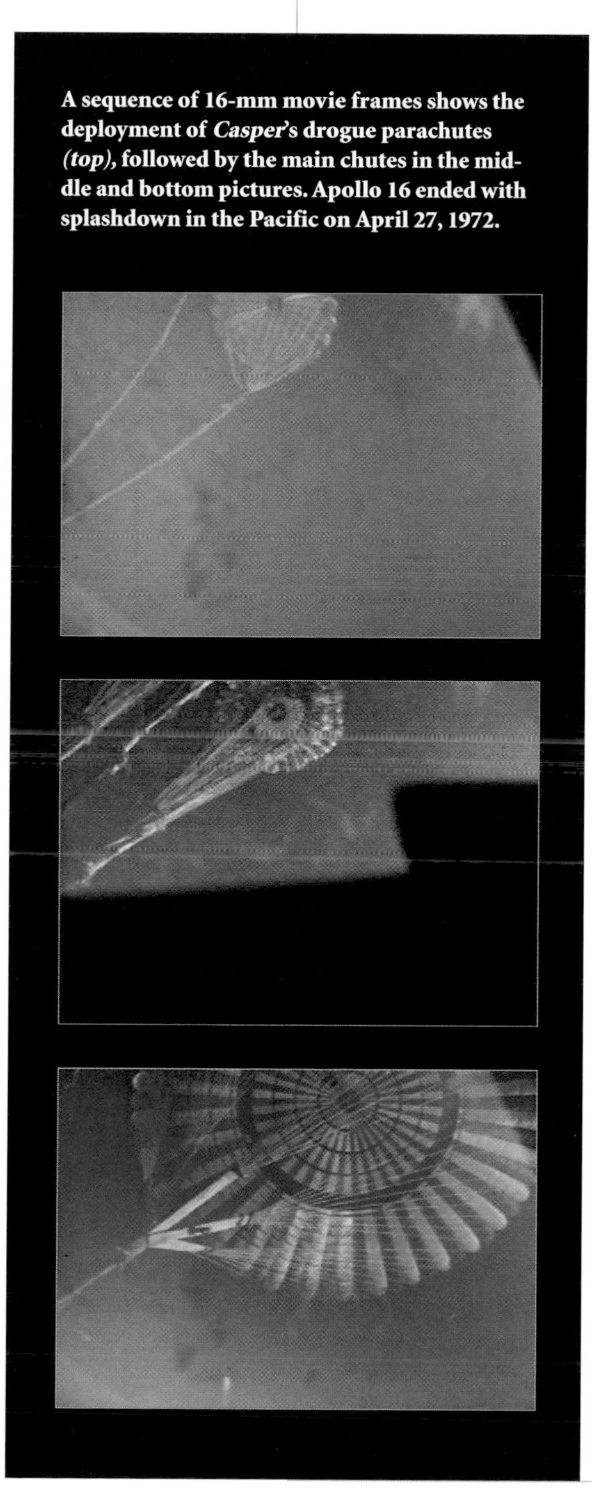

A sequence of 16-mm movie frames shows the deployment of *Casper*'s drogue parachutes *(top)*, followed by the main chutes in the middle and bottom pictures. Apollo 16 ended with splashdown in the Pacific on April 27, 1972.

"What do you think, Charlie, are they impact rocks?"

"We'll leave that to you," Duke said.

Any uncertainty vanished when the samples were unpacked in the Lunar Receiving Lab. Nearly all the rocks were breccias; the rest were ancient chunks of anorthosite. No traces of volcanic rock turned up, save for a few fragments of *mare* basalt thrown in by a distant impact. And while the Apollo 16 rocks could not rule out that lava or cinders had erupted elsewhere in the highlands, they all but proved that none had flowed at Descartes. It was not the first time the geologists had misinterpreted photographs of the moon; it would not be the last. But that did not mean there was anything inherently wrong with photogeology. The error, one geologist would write years later, was that they had neglected to define more than one working hypothesis. It was clear now that Elston and Boudette had been so enthusiastic about highland volcanism that they had overlooked alternative explanations.

But why had so many geologists hopped onto the volcanic bandwagon? In part, they had been misled by a false assumption that had plagued them since the beginning, that the moon was somehow earthlike. The resemblance of lunar highland formations to terrestrial volcanic features had reinforced the long-held idea that the moon had enjoyed a long and rich history of volcanic activity. In the wake of Apollo 16 geologists would come to understand that the overwhelming force that had shaped the surface of the moon was not volcanism, but the violence of cosmic bombardment. In the years to come, they would often recall a comment Ken Mattingly made from lunar orbit: "Well, back to the drawing boards, or wherever geologists go."

But they would also realize that Apollo 16 was perhaps the greatest leap in understanding of the moon since the first lunar landing. As one of them would later write, ". . . the mission to the Descartes Highlands illustrated once again that science advances most when its predictions prove wrong." Deciphering the evolution of the highlands would be far more difficult than anyone had imagined, because instead of a well-defined progression of lava flows, the reality was near-chaos; the origin of any particular rock or hill might be traced to a crater miles away or perhaps a giant impact basin hundreds of miles away. Two decades later, geologists would still be debating whether one of those impacts—the Imbrium cataclysm—had left its imprint at Descartes. They would search the breccias for clues to the violence that had preceded Imbrium, when huge asteroids struck the moon and earth. And in Apollo 16's collection of anorthosites, the largest of all the Apollo missions, there would be glimpses into the nature of the moon's primordial crust. The geophysicists too were rewarded when, three weeks after Young and Duke left the moon, Apollo 16's seismometer picked up the largest lunar impact ever recorded. Its reverberations showed that the crust at the Descartes highlands was more than 9 miles thicker than average. And finally, Ken Mattingly's photographs and observations, combined with the data from *Casper*'s scientific instruments, would help the scientists extend the lessons of the rocks to highland regions across the moon. Apollo 16 did more than change the image of the Descartes highlands. It reaffirmed the most basic reason for exploration. If they'd known so surely what was there, they wouldn't have needed to go to the moon to begin with. The point of going was to find out. ☾

The command ship *Casper* floats above the lunar farside as Ken Mattingly begins his three-day solo voyage in lunar orbit. This photograph was taken by Charlie Duke shortly after he and John Young undocked in the lunar module *Orion*, in preparation for their descent to the surface.

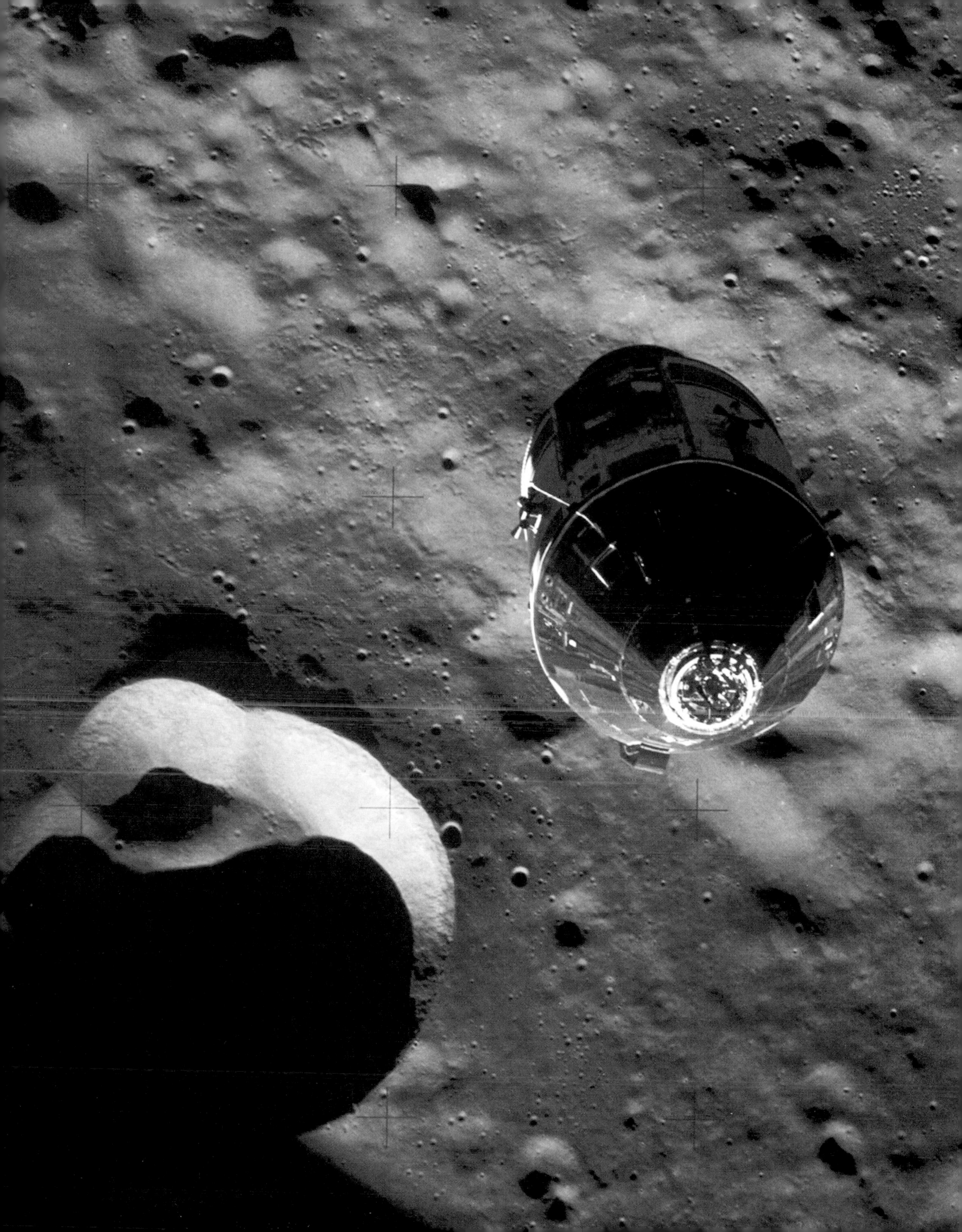

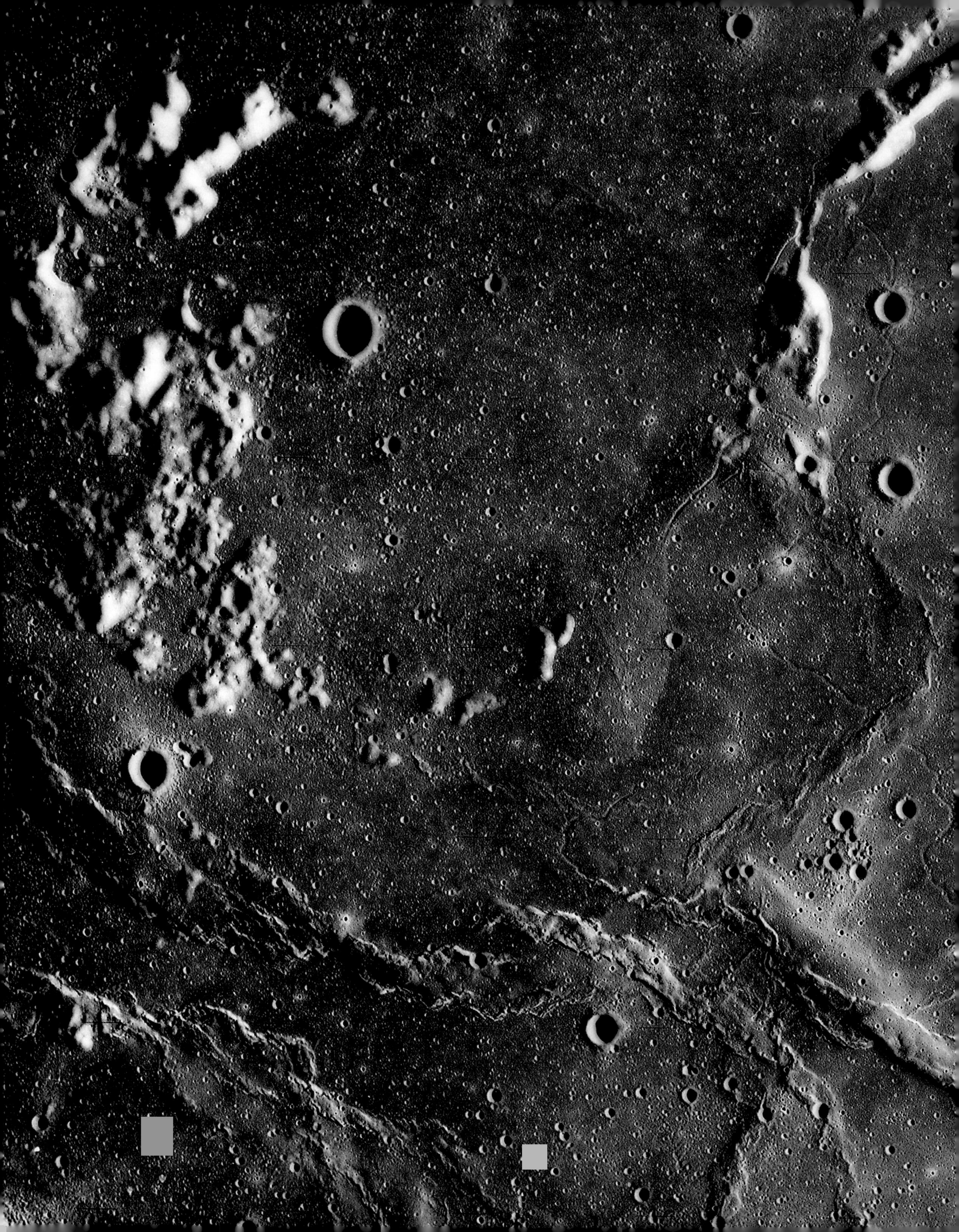

Sculpted by volcanic outpourings from the lunar interior, part of the southern region of the Ocean of Storms fills this view, which Mattingly obtained with Apollo 16's high-powered mapping camera. The broken rim of an ancient crater pokes through the lava plains at the top of the picture; ridges formed by ancient eruptions snake across the bottom of the picture.

A portrait of Theophilus crater, 68 miles in diameter, was captured in striking detail by Apollo 16's mapping camera. Among the variety of features in the photograph are the complex of central peaks *(top center)*, the hummocks of the crater walls, and the intricate pattern of ejecta surrounding the crater's rim.

The mapping camera obtained this portrait of the rugged highlands about 160 miles northwest of the Apollo 16 landing site. The two largest craters visible near the top of the frame are Halley *(left)* and Hind, each close to 18 miles across.

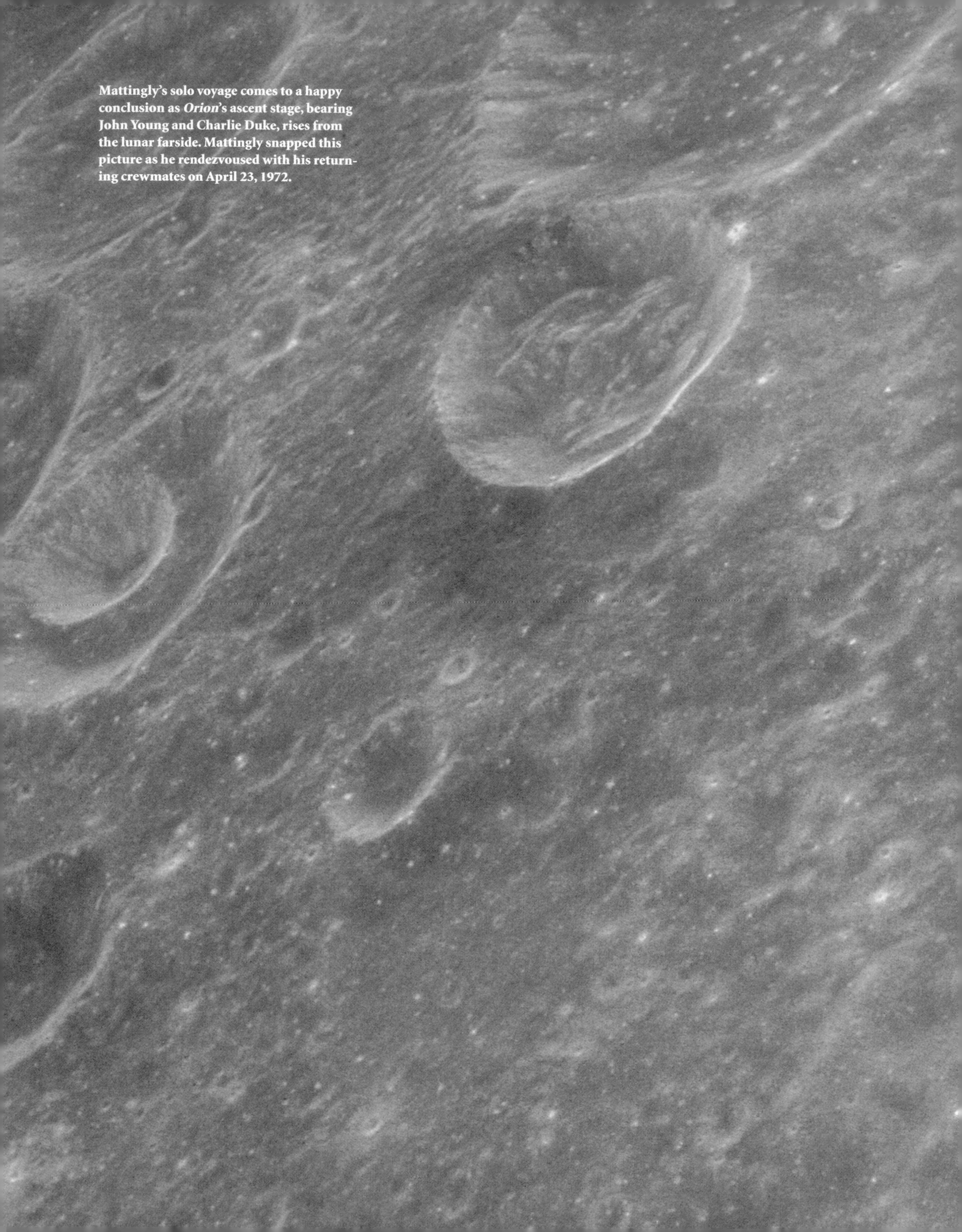

Mattingly's solo voyage comes to a happy conclusion as *Orion*'s ascent stage, bearing John Young and Charlie Duke, rises from the lunar farside. Mattingly snapped this picture as he rendezvoused with his returning crewmates on April 23, 1972.

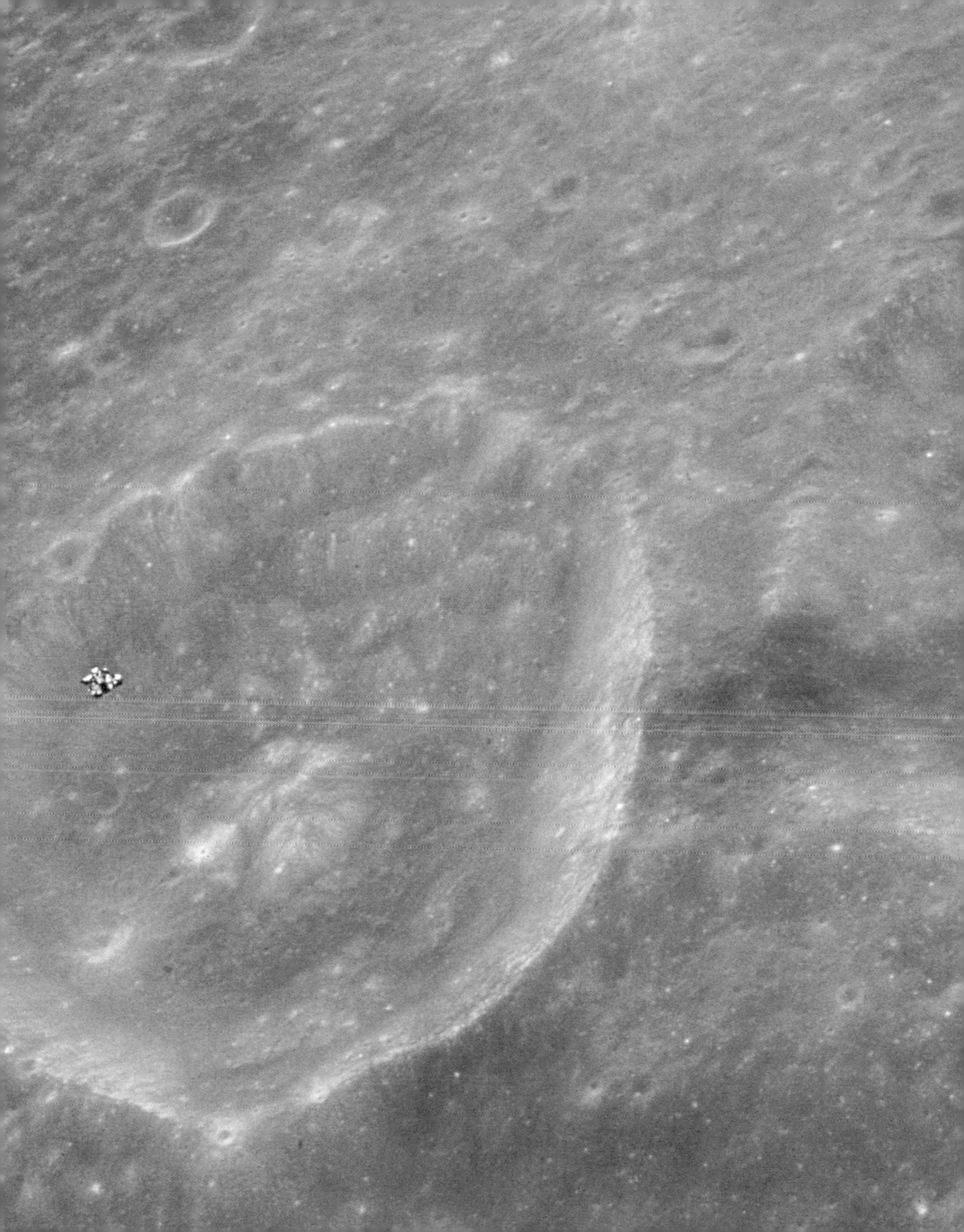

CHAPTER 13

THE LAST MEN ON THE MOON

APOLLO 17

I: SUNRISE AT MIDNIGHT

His space suit smudged with lunar dust from three days of exploration, Gene Cernan stands in the moon's Taurus-Littrow valley during Apollo 17's third moonwalk on December 13, 1972. Behind him rise the peaks of the Sculptured Hills.

Charlie Smith had seen more of the sweep of history than anyone in the United States. Born in Liberia, he'd been taken aboard a slave ship at the age of twelve and brought to Galveston, Texas, where he grew up on a white man's ranch. He'd toted a .45 since he was thirteen, and could tell tales of riding with Jesse James and Billy the Kid. In his adult life he had witnessed the invention of the telephone, the automobile, the airplane, television, the atomic bomb, and the microchip. In December 1972, Charlie Smith's age was given at one hundred and thirty years. As the oldest living American he was invited to the launch of Apollo 17, and so he and his seventy-year-old son Chester traveled from a central-Florida town called Bartow to the Kennedy Space Center. As dusk fell on December 6, they sat with dozens of other celebrities in the VIP bleachers near the Vehicle Assembly Building. Across the waters of the tidal basin the last of the moonships gleamed in the floodlights, transformed into something mythic. The light from Pad 39-A could be seen all around Brevard County, in Cocoa, and Eau Gallie, and Titusville, towns for whom the imminent departure of a team of moon voyagers had become almost commonplace. Only in the

Former slave Charlie Smith *(center)*—at age 130 the nation's oldest citizen—attended the launch of Apollo 17 with his 70-year-old son Chester *(right)*. Smith had witnessed technical advances for well over a century, but he found it impossible nonetheless to believe that human beings were voyaging to the moon.

first decades of this century—when Charlie Smith was in his seventies, his eighties—had the exploration of this planet marked its last major objectives: the polar crossings, the Himalayan treks. As the scope of exploration had broadened, the pace of time had compressed. Less than twelve years had elapsed since Kennedy's decision to go to the moon, just four years since Apollo 8 lifted off from Pad 39-A. Now, at the same launch pad, the first Age of Lunar Exploration would enter its last act. ☾

By now, of course, that was old news. In Houston, the space center was already gearing up for the Skylab flights that would begin in the spring, and beyond that was the joint mission with the Soviets, the Apollo-Soyuz Test Project. And then, in the late seventies if all went according to plan, would come the Space Shuttle. It tugged at the flight controllers even as they prepared for the last moon mission. It would launch like a rocket and land like an airplane. It would withstand the fires of reentry unscathed and be ready to fly again in a matter of weeks. It would be a technological marvel; it would be a true flying machine. It would have *wings.* Now was the time to move away from Apollo's throwaway lineage and make space economical, and that was the shuttle's main billing: the first reusable spacecraft. By the 1980s, NASA hoped, the shuttle would make getting to space so routine that the space program would turn a profit. It was nothing less than the future of spaceflight.

Most of the Astronaut Office was already working on it. And other than the three men in that rocket, and their backup crew, there were no veterans of lunar missions who were still working on Apollo. Eleven of them had left the astronaut corps to begin Life After the Moon, with varying results. And then there was Deke Slayton. After ten years, he was finally back on flight status. He'd undergone a risky procedure at the Mayo Clinic in which a catheter was inserted through his veins and into his heart, so that the doctors could see once and for all whether it had been damaged—and he passed with flying colors, literally. There was talk he might be assigned to fly on the joint Soviet-American flight in 1975, by which time he would be fifty-one years old. He was already studying Russian. ☾

But on the whole, it had not been a good year for astronauts. In the spring of 1972, it had come out that Dave Scott and his crew had carried four hundred unauthorized first-day covers to the moon and back. The men kept three hundred of the covers and gave the remaining hundred to a German stamp dealer named Horst Eiermann, with the understanding that they would be sold to collectors after the Apollo program was over, privately and with no publicity, with the three astronauts sharing equally in the profits. Each man stood to gain $8,000 by the deal, with which they planned to set up trust funds for their children. But Eiermann began selling the covers within weeks after Apollo 15 returned. Scott's crew immediately notified Eiermann that he had broken the terms of the agreement, that the deal was off. All three men refused to take any money. But word of the sales got into the European press, and it was only a matter of time before NASA found out. By the fall there were reports that envelopes were selling for $1,500 apiece. When rumors began circulating, Deke Slayton went out on a limb to defend his people, saying he didn't believe the stories. Then Slayton found out, from Jim Irwin, that they were true. He was furious. Congress was demanding an investigation into improper conduct by astronauts. NASA formally reprimanded Scott and his crew. By June, Irwin had announced that he was leaving NASA and retiring from the air force to launch a Baptist ministry called High Flight. Al Worden was transferred to NASA's Ames Research Center near San Francisco. And Dave Scott was moved to a management post within the Manned Spacecraft Center.

The other astronauts were divided in their reactions. Some saw it as simply a dumb mistake. Others thought Scott, as crew commander, should be court-martialed. To some, it was a gray area. Astronauts had sold their autographs, for example, and profited in other less dramatic ways from their fame. But this time, there was so much money involved, and it had all become so public. It had tarnished the astronaut corps. That it had all been done by earnest, straight-arrow Dave Scott, whose mission had been such a high point for Apollo, only made the shock greater. Years later, Scott would blame his bad judgment on the pressures of getting ready for a lunar landing mission.

Whatever the astronauts thought of the stamp affair, the damage was done. Within NASA, the people who had always felt the Astronaut Office had too much clout were determined to see their wings clipped. For better or for worse, the myth of the Perfect Astronaut had crumbled. Now the public knew that astronauts didn't always follow the rules. Astronauts were fallible; they were human. ☾

By December, the stamp affair was old news. In the VIP stands, the atmosphere was something like the last performance of a long-running stage play. Some of the stars of that production—several of the men who had crossed the translunar gulf, including Jim Lovell, Neil Armstrong, and, with special irony, Dick Gordon—were on hand to see the last of their comrades

"I see that's a rocket, but th' ain't nobody goin' t' no moon. Me, you, or anybody else."
— Charlie Smith

leave the earth. Nearby, more than a thousand newspeople filled the press site. And a crowd estimated at more than a million people jammed the roads and beaches. All of them were lured by the promise of a space spectacular: The particulars of Apollo 17's launch window required that this be the first—and in fact, the only—night launch of the Saturn V. And then there was Charlie Smith, to whom this whole thing was literally beyond belief. Wearing a Stetson and a string tie, and an Apollo 17 mission patch on his lapel, Smith gazed at the Saturn and said in a sly but paper-thin voice, "I see that's a rocket, but th' ain't nobody goin' t' no moon. Me, you, or anybody else."

And for a while, that night, it seemed that the old man was right. As the launch time of 9:53 P.M. approached, lightning danced among distant purple clouds, seemingly ready to threaten the launch but never doing so. Behind the party-din of conversation the loudspeakers carried the voice of the public information officer, Jack King, his flat monotone counting down the last hours and minutes. Vapor issued from the Saturn, its stages filling with super-cold propellants. Now the crowd quieted and the voice on the loudspeakers fully emerged, echoing in the calm, damp night, counting down, now five minutes to go, now one. Then, at thirty seconds, the steady cadence of numbers halted. "We have a cutoff," said the voice. "We have a cutoff."

Three and a half miles away, at the pinnacle of the rocket, Gene Cernan, Ron Evans, and Jack Schmitt waited tensely inside the command module *America,* all thoughts on the abort handle in Cernan's left hand. No other moment could have totally crystallized the responsibility that went along with command of a spacecraft and a $450 million lunar mission. Cernan had trained for Apollo 17 as the climax of his career; he was not going to be denied. If the Saturn's guidance system went out during the launch, he was prepared to fly it into orbit himself, using the 8-ball, the hand controller, and the stars—but let this machine get off the ground. Don't make Apollo 17 end with a twist of the abort handle.

There was no danger. The launch pad's automatic sequencer, which controlled each split-second event in the complex launch sequence, had failed to pressurize the Saturn's third stage. In the blockhouse 3½ miles away, the launch controllers noticed the problem and sent the necessary command directly to the booster. But they were too late. The sequencer, aware of its own error, didn't accept their action as a substitute; it stopped the count on its own. ☾

For a time it seemed depressingly likely to Cernan and his crew that they would not launch that night, but soon there was welcome news. The launch teams would work around the malfunction; it would take a couple of hours. During the delay Evans fell asleep, and Cernan and Schmitt got on the air to complain about the snoring.

Far from the crowds and the VIPs, at the special viewing site reserved for astronauts, their families and friends, Jan Evans waited tensely with her children, Jon and Jaime. Barbara Cernan and her daughter, Tracy, were here too. All day Jan had been going on pure excitement; it had to be that way after all those parties, like the bash thrown by *Life* magazine, and the other celebrations that had been in full swing for days. The tension she felt now was not out of fear; Jan had too much faith in the hardware for that. It was out of the desire to see her husband do what he had longed to for the past six years. From this same grassy shore, she and Ron had watched most of the moon flights begin. They had heard thunder and shed tears, out of elation for their departing friends. And Ron had always said, "One day, that's going to be me."

IN 1972

At the 20th Summer Olympic Games in Munich, Mark Spitz wins or shares seven gold medals for swimming. Eight Palestinian terrorists invade the Olympic Village and murder two Israeli athletes.

The last U.S. ground troops leave Vietnam.

Billy Jean King wins both the U.S. Open and the Wimbledon women's singles tennis championships.

Aboard the aircraft carrier U.S.S. *Ticonderoga* in 1966, fighter pilot Ron Evans *(left)*, who has just returned from flying a mission over North Vietnam, hears from his commanding officer that he has been selected as an astronaut. Evans came to NASA as part of the astronaut group who called themselves the Original 19.

Jan was as devoted to her husband as any astronaut wife. In 1965, when she and Ron had a house near Miramar Naval Air Station in San Diego, a letter arrived informing him that he had the qualifications to apply for the fifth group of astronauts, and asking whether he wished to volunteer. At that moment Ron was 8,000 miles away, flying sorties from the U.S.S. *Ticonderoga* off the coast of North Vietnam. The letter said all applications must be in within ten days. Jan sprang into action. She worked up her courage and called Deke Slayton herself, and told him, "Ron *definitely* volunteers." Slayton assured her they'd accept his application late.

Many months later, after he had gone home for the physicals and the interviews, and then rejoined his ship, Evans returned from a mission and got the message to report to the ready room. Wondering what he had done wrong, Evans arrived to find the captain and several of the other pilots gathered, and the captain read the letter informing Ron that he had been selected as an astronaut. His shipmates would say later, he just about floated out of his chair.

Hours ago, just before nightfall, Jan and the children had waited at the entrance to the crew quarters to say good-bye to Ron. First came Gene Cernan, emerging into the glare of the television lights and blowing a kiss to nine-year-old Tracy. Ron was right behind him, grinning, giving a little skip as he came out, like a kid. For a moment he stopped, and she put an arm around the bulk of his white space suit and planted a kiss on the clear bubble helmet, and watched him climb into the transfer van, followed by Jack Schmitt. Then the van headed off into the night.

Now, for Jan Evans, there was only waiting. It was past midnight now. Just as the delay seemed interminable, the count entered its final minutes, heading for a launch time of 12:33 A.M. Once more the voice on the loudspeaker counted down, and this time, it did not stop. At the 8-second mark there was an explosion of red flame underneath the Saturn that gave way to a river of incandescent white. The sight of it drew from the crowd one great *Ahhhhh!* as twin plumes of fire and yellow smoke issued into the dark sky on either side of the launch tower. At the moment of release, as the great rocket rose from the earth, the sky filled with a brilliant golden light, like the light of sunrise. As the crowd cheered, the announcer's voice, suddenly electric, was heard, "It's lighting up the sky, it's just like daylight here at the Kennedy Space Center . . ."

Headed for the launch pad, Evans attempts a farewell kiss for his wife, Jan, despite his bubble helmet. During a delay in the ensuing countdown, Evans fell asleep in the command module, eliciting complaints about his snoring from his crewmates.

Then the noise hit. Up to now no one had been aware of the eerie silence, as the shock waves rolled across the tidal basin, arriving at the viewing sites just as the Saturn cleared the launch tower. It was a rippling, ear-splitting staccato roar that pummelled the chests of the surprised spectators and shook the ground. It rattled houses all up and down the Cape. Far beyond, the glow from this second sun was visible across Florida and as far away as North Carolina. Jan Evans stood transfixed by the rocket carrying her husband, never taking her eyes off it, not even when she was suddenly aware of a great commotion in the water in front of her. Out of the corner of her eye she saw fish jumping into the golden light, so many that the water seemed to boil with their thrashing until, at

Turning midnight into dawn, Apollo 17 thunders aloft on December 7, 1972, the only night launch of the Saturn V moon rocket. Former astronauts, celebrities, and more than a million spectators were on hand for the event. More distant observers reported that the light from the ascending rocket was visible for hundreds of miles.

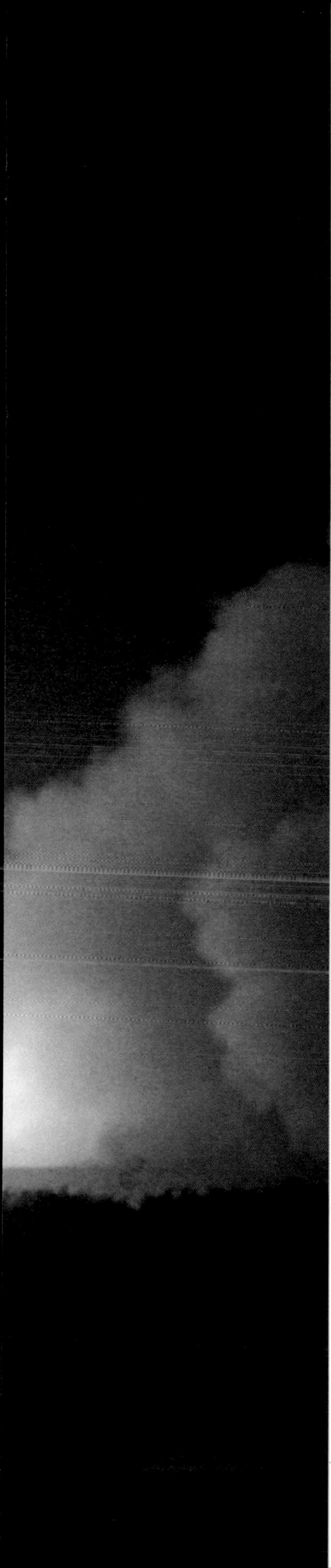

last, the sonic wave crested. The rocket was now a glowing torch suspended high overhead, its flame fanning out in the rarefied air at the outer reaches of the atmosphere.

Now there was stillness again. She could hear Gene Cernan's voice, charged with adrenaline, echoing from the loudspeaker in curt exchanges with Houston. When the first stage fell away there was a sudden, brilliant flash. A moment later the escape tower departed and she heard a cry of delight from Ron: "*Aaa-ha!* There she goes!" But that was all; from now on she heard only the clipped, technical jargon of the mission. She did not cry, not this time; she was possessed of a wonderful, joyous calm. As she watched her husband's rocket dwindle to a bright star she felt that a part of her was going with him to share the adventure for which he had worked so long. He was way out over the Atlantic now, hundreds of miles away from her already. She strained to follow Apollo 17 into the night with her eyes, but before it reached orbit, it was gone.

Meanwhile, at the VIP bleachers, Charlie Smith, who had watched this extraordinary leaving with a steady gaze, said, "I see they goin' somewhere, but that don't mean nothin'."

●●◐○○○◑●

Nine years of a man's life can go by in a flash; at least it sometimes seemed that way to Gene Cernan. In 1963, the year he and Barbara went to Houston —he couldn't believe he had been chosen over so many pilots with test pilot credentials—their daughter, Tracy, was born. When he flew Gemini 9 she was only three; she didn't understand what was happening. By the time he went to the moon on Apollo 10, she was six—old enough for him to take her outside one night a month before launch and try to explain things: The moon was very far away, where God is, and hardly anybody had ever been there, but he and Mr. Stafford and Mr. Young were going there. And when he came back, he took Tracy out under the moon once more, and she said, "Daddy, you went to the moon! And the moon is way, far away, where God is—and you went with Mr. Stafford and Mr. Young, and hardly anybody else has ever been there." Then she got quiet for a moment, and Cernan figured she was about to plumb the depths of six-year-old curiosity, and that his answer would be the makings of tomorrow's show-and-tell. What he forgot was that for a little girl whose next-door neighbors were astronauts, going to the moon wasn't a very big deal.

"Daddy?"

"What, Punk?"

Tracy Cernan *(left)* and Amy Bean proudly display photographs of their moon-voyager fathers. On the wall above Tracy are photos from Gene Cernan's Gemini 9 mission.

"Now that you've been to the moon, when are you going to take me camping like you promised?"

Tracy Cernan was nine now, and she understood very well that her daddy was going back to the moon, and that this time he was going to walk on it. Probably, though, she didn't realize how much it meant that he was returning as a mission commander. That was even more important to Cernan than crossing the last 50,000 feet he had missed on Apollo 10; it was so important, in fact, that Cernan had actually turned down a chance to walk on the moon as John Young's lunar module pilot. It had been a calculated gamble; Slayton had not promised him another flight after Apollo 10, and even after he was assigned to back up 14, there was absolutely no guarantee that he and his crew would get their own flight. But in August 1971, his gamble had paid off. And if command was the greatest challenge he had faced in his career, then Cernan had grown into it. During his first two flights, Gemini 9 and Apollo 10, Cernan had been Tom Stafford's shadow; though he treated his crew like equals, Stafford was a natural leader and a strong commander who knew how to use his power. Cernan had learned much simply by being around him. And looking back on Apollo 10, he'd come to realize how little he knew then about his spacecraft, when he thought he had known so much. Now, at age thirty-eight, Cernan had matured into a seasoned astronaut.

In a sense, he was an amalgam of the broad spectrum of personalities in the Astronaut Office. Like most of his colleagues, his ego was among the healthiest in the country, but those who met him were always pleased to find him "just a regular guy." Cernan was one of the astronauts whom people who worked at the space center liked to introduce to their friends. Here was this charismatic six-foot figure with cool blue eyes, salt-and-pepper hair, and quiet poise—like a good PR man, Cernan could make a stranger feel immediately comfortable. He seemed at once to savor the hero's mantle and to disown it. To be sure, he enjoyed the social perks that went along with the job; he made the Houston party scene with as much enthusiasm as Wally Schirra and Jim Lovell ever had. Richard Nixon's vice president, Spiro Agnew, played golf with him and dined at his home. He and Barbara had the aura of a storybook couple.

If Cernan had stayed in the navy, his goal would have been to be a squadron commander, and with the Apollo 14 backup crew he had his squadron. It even had the squadron's spirit—what other team had come up with its own backup patch? They were what Ron Evans called a "crew-crew." They flew together; they played together. And they would've gone to the moon together, had it not been for the cancellation of Apollo 18. Cernan had seen that coming, he understood that NASA would have been foolish to have a geologist available for the last landing and not send him—especially considering the storm of criticism that would have unleashed. Still, Joe Engle was his friend and his crewman, and Cernan had fought to keep him on, all the way to NASA Headquarters. But it had come down to an ultimatum: accept the change or lose the mission. And in that episode, Cernan had learned the limits of the spacecraft commander's power.

Now that was ancient history. Cernan would be the first to acknowledge that Schmitt had earned his seat. He knew that many of the Old Heads thought putting a scientist on a lunar mission was too risky, that NASA had caved in to pressure from the scientific community, and that it was a mistake. Cernan didn't agree. Schmitt was not the greatest aviator the world had ever seen, but certainly adequate, and during the flight he wouldn't do any flying anyway. More important, he was a damn good lunar module pilot. His geologic expertise, of course, was unquestioned. As to whether Schmitt would really be able to *use* that expertise on the moon, working to the timeline and encased in a pressurized space suit—that was something Cernan, and many astronauts, wondered about. But he was certain of one thing: he and Schmitt would stay on the moon the longest, collect the most rocks, take the most photographs, bring home the most data. They would make the last the best.

NASA's managers did not seem to share his spirit. To them, Apollo 17 was like a last-minute, all-or-nothing bet at the blackjack table—because if Cernan and his crew didn't come back from the moon, the shuttle and the rest of the space program would be in jeopardy. Some confided to Cernan, "If it were up to me I'd cancel this flight." Even Chris Kraft, who had succeeded Bob Gilruth as director of the Manned Spacecraft Center, was telling him, "Take off your white scarf. We don't want to lose anyone now. Don't take any chances out there; just get home alive." Understandable though it was, this

The Apollo 17 astronauts were the first to see the earth fully illuminated, shortly after beginning their moonward voyage. Africa and the Arabian Peninsula dominate this view; the Antarctic icecap is visible at the bottom.

kind of talk rankled Cernan. He sensed, as Neil Armstrong had in the months before Apollo 11, the damage that failure would cause the program and the national image. And as a mission commander, he was already conservative; he didn't need to hear Kraft telling him not to take unnecessary risks.

Whatever the mood in Houston, Cernan was concerned that elsewhere, Apollo's imminent end was hurting morale. It was true that at the Cape, where technicians readied his Saturn V for launch, the biggest job cuts were long past, but nine hundred more workers were due to be laid off after the mission. In Downey, California, the Rockwell teams were at work on command modules for the Skylab missions and the joint Soviet-American flight. But on Long Island, the Grumman people were literally working themselves out of a job; the entire lunar module program would be over when Apollo 17 left the moon. Cernan and Schmitt spent a lot of time at Grumman, and if morale was bad, they never saw it. It amazed Cernan how they worked as hard, if not harder, on the final lander; maybe he and Schmitt had something to do with that.

At the Cape and in Houston, Cernan's crew organized softball games with flight controllers and support teams. They threw parties. Years later, those who were involved would remember this Apollo crew for those times. They would also remember that Cernan, who had never been shy about public speaking, was in his element. Sometimes after a few drinks—and sometimes not even after a few drinks—he would stand up on a chair and launch into an oration that was part pep talk and part campaign speech. "Apollo 17 may be the last flight to the moon," Cernan would say, "but it's not the end. It's the end of the beginning." And for the most part, it worked, though some of his audience smiled to themselves at this mixture of ham and sincerity. Later they teased, "Gene, I'm not sure I've got it straight—is it the end or the beginning?"

●●◐○○○◑●

Whatever you called it, Apollo 17 was the last roll of the dice for the geologists; only one chance remained to unravel the knotted ball of string called lunar evolution. Before the flight, when the press queried him on the value of yet another moon mission, Jack Schmitt liked to say that the moon had as much land area as the entire continent of Africa—and yet, astronauts had explored only a handful of acres. How could anyone think that an entire world might give up its secrets after only a half-dozen modest expeditions? Obviously, the choice of landing site for Apollo 17 was especially consequential, and when he was named to that mission Jack Schmitt had resurrected his goal of landing on the far side, on the dark lava floor of the crater

LM 12
DESIGNED AND BUILT BY
A TEAM OF EXPERTS FOR
GENE CERNAN
AND
JOE ENGLE
FLY IT IN GOOD HEALTH
FROM
THE LM 12 TEAM

Some of the Grumman workers who built the Apollo lunar modules gather around LM number 12—the final moon lander—to bid good-bye to their creation in June 1971, just before the craft was shipped to Cape Kennedy. The workers wore color-coded hats that represented specialties such as propulsion and environmental control.

Tsiolkovsky. It was a dream site, with not only a far side *mare* to sample—would it differ chemically from those on the near side?—but a gigantic central peak that offered samples from deep within the far-side crust. Needless to say, Schmitt didn't need to sell the idea to the geologists. But when he talked it up to his flight controller friends they half-jokingly responded, "Well, Jack, out of sight, out of mind." There would be no earth in the sky above Tsiolkovsky, and the only way to communicate would be via special relay satellites in lunar orbit. A pair of Tiros weather satellites were already available as off-the-shelf spacecraft; a single Titan booster could place both of them at a point about 30,000 miles above the far side—one of the so-called libration points—where gravitational conditions were such that they would remain suspended over the surface, within sight of both Tsiolkovsky and earth.

Schmitt had first raised the idea late in the spring of 1970, as part of an effort to save Apollo 18 and 19: if he could show that those missions could reach places where the geologists had always wanted to go, maybe NASA would think twice about canceling them. But at the space center, it seemed everyone was too caught up in the recovery from Apollo 13 to consider such schemes. Now, in the fall of 1971, Schmitt pushed for Tsiolkovsky again—this time, to generate renewed public excitement for Apollo. But it was no easier to sell the far side now than it had been the year before. The satellites added a new level of complexity to an already risky enterprise. Most of all, there was no money for them. Cernan discussed the idea with George Low and Chris Kraft and was told it wasn't going to happen, but Schmitt kept pushing until Kraft told Schmitt to stop talking about it. Tsiolkovsky's secrets would have to wait for the next generation of lunar explorers. ☾

No matter; the geologists had found a place that could only be described as the jewel in the crown of Apollo landing sites. It gained this status during Apollo 15, when Al Worden flew over the southeastern shore of Mare Serenitatis (the Sea of Serenity, which forms the Man in the Moon's left eye) and saw "a whole field of small cinder cones down there." His verbal pictures of tiny volcanic craters—whose conical shape seemed a dead giveaway—were a summons to exploration. But the real clincher came when the film from Apollo 15's panoramic camera was unreeled on light tables at the space center, and the geologists could see with their own eyes what had so excited the lone astronaut. Some of those craters lay within a small box-canyon ringed by the steep-sided peaks called the Taurus Mountains that formed the worn, ill-defined rim of the ancient basin. Seventeen miles to the northeast lay the crater Littrow. What was so striking about the valley of Taurus-Littrow was its floor, which was covered by some of the darkest material seen on the moon. It seemed entirely out of place, appearing as it did among the light-colored highland terrain. Whatever it was, it was widespread; the geologists found it all along the southeastern portion of Mare Serenitatis. They even spotted patches of it on the high elevations of some of the mountains. How had it gotten there? On earth, that can happen when pockets of volcanic gas suddenly find release at the surface, spraying molten rock high into the sky. If the same thing had happened on a very intense scale on the moon, it could have blanketed a wide area with volcanic ash. The coating probably wasn't very thick; on Worden's pictures the rims of small impact craters were clearly visible poking up through the dark mantle. Because it was so dark, the geologists surmised the deposit hadn't been exposed for very long to micrometeorite rain; otherwise it would have become faded by being mixed with the soil. Perhaps it was only half a billion years old. After the disappointment of the Descartes highlands, here was new reason to hope that the moon hadn't died geologically so long ago after all.

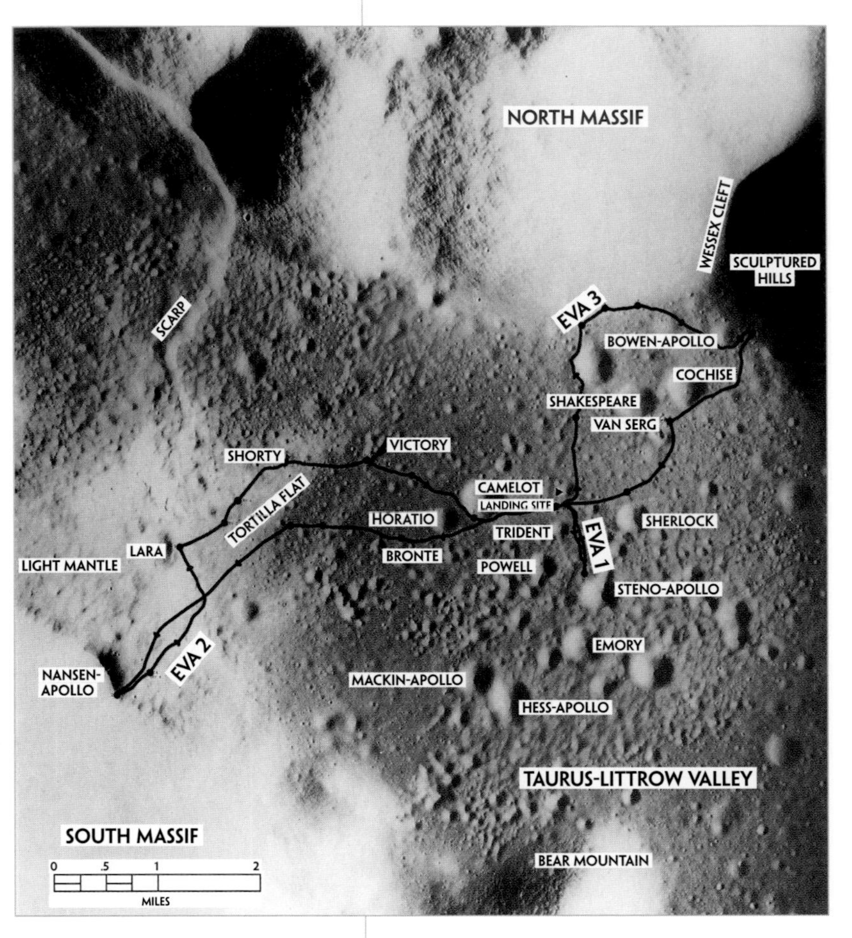

Apollo 17's three moonwalks in the Taurus-Littrow valley included visits to two mountains, called the North and South Massifs, and to a number of craters. The astronauts named many of the features for explorers, writers, poets, and characters from literature.

At the same time, the valley walls—and in particular, two rocky prominences christened the North and South Massifs—might finally give a look back to the time before the Imbrium impact, whose influence had yet to be fully escaped. Everyone agreed that the Serenitatis basin had formed before Imbrium, and hoped the Massifs would contain the oldest lunar rocks yet found. And there were still other lures: a probable landslide at the base of the South Massif that would undoubtedly contain a rich variety of samples from

The command ship *America* appears as a tiny miniature set against the North Massif and the Apollo 17 landing site. Gene Cernan snapped this picture soon after undocking, as he and Jack Schmitt prepared for their descent to the moon aboard the lunar module *Challenger*.

the mountain; a scarp that resembled features called wrinkle ridges commonly seen on the *maria* but never visited; and the Sculptured Hills, whose knobby shape set them apart from the Massifs and sparked renewed hopes of highland volcanism. Geologically speaking, Taurus-Littrow was the most complex site yet, worthy of the full capabilities of the J-mission. Here, in one place, lay the possibility of exploring the beginning and the end of lunar history.

But getting there wouldn't be easy, primarily because the valley was only 4½ miles across, from one rocky wall to the other. The normal target ellipses—the ones that showed possible trajectory errors—spilled up onto a mountain on one side of the valley and a landslide at the other. But that was because the ellipses were very conservative. Perhaps the engineers had never gotten over the fact that Armstrong and Aldrin had landed 5 miles past their aim point, because the planners had never changed the size of the error ellipse, even though the next four lunar modules had set down within a few hundred feet of their targets. One of the men involved in the site selection process, Jack Sevier, made that point to the trajectory experts, and they soon came back with a smaller ellipse that fit neatly within the valley. Now it was up to the managers. At the site selection meeting, Kraft and McDivitt flashed a thumbs-up, as if to say, "It's the last one; let's go for broke."

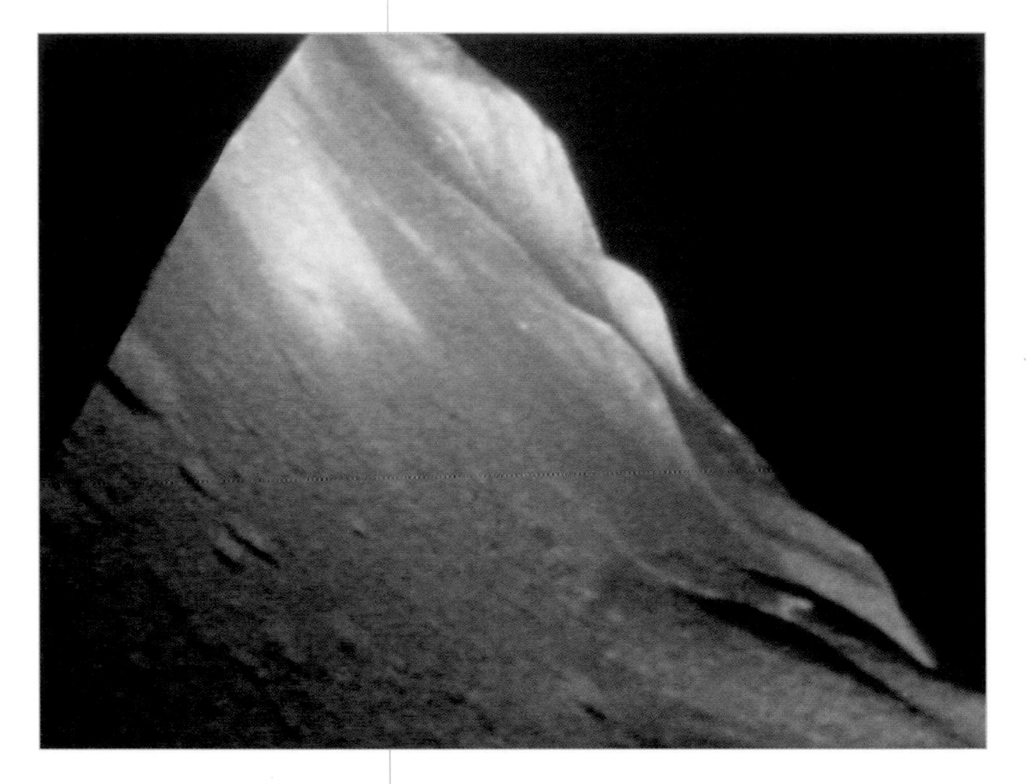

***Challenger*'s onboard 16-mm movie camera recorded this view of the Taurus-Littrow valley as the lander pitched forward for its final descent. The large crater in the upper left part of the frame is Camelot, 2,360 feet across.**

In part, the geologists chose Taurus-Littrow because they knew Jack Schmitt would be there to explore it. But before he and Cernan could turn their combined geologic prowess on the valley, they had to get there. Schmitt had no unease about becoming the first of his chosen profession to practice on another world, but he knew he must not turn in anything less than an outstanding performance as a lunar module pilot—not just for the sake of the mission, or the lives of his crewmates, but because he was nothing less than a test case for all the scientist-astronauts who were waiting for their chances to fly. Before he could be a geologist, Schmitt would have to be an astronaut. And on the afternoon of December 11, when he and Cernan stood side by side at the controls of the lunar module *Challenger,* descending into the valley of Taurus-Littrow, the transmissions that came down on the earth-moon airwaves were the same jargon-rich shorthand that had been the soundtrack to each of the five previous lunar landings. Schmitt's concentration was total. Except for a glance at the moment of pitchover, he did not look outside. He did not see the dark valley opening before them, its white mountain sentinels casting long, pointed shadows. He did not see the smooth place, just near the crater Poppy (Tracy Cernan's nickname for her late grandfather) where Cernan planned to set down. When he heard Cernan say, "I don't need the numbers anymore," Schmitt knew his commander was getting ready to take over from the computer, but he did not know that there was a boulder field out ahead that Cernan had to maneuver around. At last, 60 feet above the moon, Schmitt took another brief look and reported, for the benefit of his listeners on earth, "Getting a little dust. . . . Very little dust." Then he returned to the instruments, unaware that *Challenger*'s spindly shadow had appeared on the dark plain, moving toward them, now passing under them. "Stand by for touchdown," Cernan called, and *Challenger* crept down the last 25 feet; as soon as Schmitt saw the blue contact light come on he called it out, and after a moment of free fall, he felt them come to a stop with a firm thud. Even as Cernan took a moment to exult, "Okay, Houston, the *Challenger* has landed!" Schmitt was reading checklists and flipping switches. Later, he would jokingly complain he had missed the whole landing.

MONDAY,
DECEMBER 11, 1972
6:08 P.M., HOUSTON TIME
4 DAYS, 21 HOURS,
15 MINUTES MISSION
ELAPSED TIME

"Hey, who's been tracking up my lunar surface?" The answer to Jack Schmitt's question was Gene Cernan; now, with characteristic irreverence, Jack Schmitt was climbing down the ladder to make his own tracks in ancient dust. And when he was at last standing on the valley floor—the end of a journey that began in Flagstaff eight years earlier—his thoughts were on the significance of the moment, not to himself but to humankind. What had brought Schmitt to NASA, and had fueled his fanatical drive toward lunar exploration, was his devotion to history. He didn't need to be told that he and Cernan would only scratch the surface, that the real exploration of the moon would be left to the ones who would follow them here. How many years or even decades away that was, Schmitt had no idea; he knew only that the space

Gene Cernan takes Apollo 17's Rover for a test drive early in the mission's first moonwalk. Seen here in its stripped-down state, the Rover will soon be loaded with a TV camera, communications antennas, geology tools, and other gear.

program was in serious trouble. But Schmitt had a kind of stubborn optimism; he would not give up hope.

"Boy," said Cernan, "your feet look like you just . . ."

Schmitt finished the thought. "Walked on the moon? Well, I tell you, Gene, I think the next generation ought to accept this as a challenge. Let's see them leave footsteps like these some day." And for a moment, as he took his first steps, Schmitt's gaze went to a landscape unlike any he had ever seen on earth. The Massifs rose like flattened pyramids into the velvet sky, their sides impossibly steep. They had neither the hewn-granite sharpness of the Rockies, nor the smooth, glacial roundness of Norway's fjords, where Schmitt had done the research for his doctoral thesis. He was looking at mountains that dated back almost to the formation of the solar system. Out on the dark plain, he saw rolling hills, their fluid forms littered with boulders ejected from the larger craters. In the near field he noticed small, rimless hollows in the soil, the center of each miniature crater marked by a dark spot of fused glass. At his feet, soil and broken rocks sparkled in the unfiltered sunlight, their textures and subtle hues amazingly vivid, their crystals iridescent. On the surface of the larger cobbles he saw small white spots where micromete-

Two huge mountains, South Massif on the left and North Massif on the right, define the Taurus-Littrow valley where *Challenger* came to rest on December 11, 1972. Jack Schmitt took the pictures that this panoramic view comprises early in the first moonwalk, as Gene Cernan powered up the Lunar Rover.

orites had blasted tiny craters. Everywhere he looked, on every scale, his eyes found new detail. He said, "A geologist's paradise if I ever saw one." Around him, light reflected from the distant mountains and nearby slopes penetrated the shadows. In the southwest, his gaze was arrested by a sparkling blue gibbous earth, hanging over the South Massif. He said nothing.

6:41 P.M.

It was only natural that Gene Cernan had come to know Jack Schmitt well; the two men had been living in each other's pockets for sixteen months. Cernan felt a kind of fatherly responsibility for him. He'd gotten used to Schmitt's outspoken manner, and his penchant for going to management with a pet issue; like Dick Gordon, Cernan had learned to keep Schmitt from getting out of hand. They'd developed a great working relationship, in the simulator and in the field. But Cernan could not say that he really knew what made Schmitt tick. Schmitt was hard to get at. As outspoken as he was, he kept his emotions well guarded. Cernan was probably one of the most

gregarious astronauts, but Schmitt was undoubtedly one of the most private. Before simulator runs, Cernan and Evans would be gabbing with the instructors; Schmitt would be off in a corner with the training manuals.

Cernan could certainly tell that Schmitt was excited to be here. In between his expert descriptions—"The basic bright-colored rock type in the area looks very much like the cristobalite gabbros in the *mare* basalt suite . . ." —Schmitt was broadcasting his excitement in snatches of song (selections included "Bury me not on the lone prairie," "What is this crazy thing called love," and "We're off to see the wizard") and horrible puns. Informed by Capcom Bob Parker that his suit temperature was running a little high, Schmitt responded, "I'm just a hot geologist; that's all."

But underneath all that brashness, was Schmitt absorbing this experience on a personal level? During the past several days, Cernan had hoped Schmitt and Evans were taking it all in for themselves. The night launch was a sight to behold—even before they lifted off, as he could feel the Saturn V's engines coming to full power, Cernan glanced out through the little window in the boost protective cover and saw the glow—and though only he could see it, he said, *"Look at the light."* In earth orbit, knowing they would be leaving only too soon, he'd made sure his crew saw the sunrise. And yesterday, just before the Lunar Orbit Insertion burn, Cernan had glanced out the command module's hatch window and had seen what he had never witnessed on Apollo 10: an enormous crescent moon bathed in sunlight, only 8,000 miles away and growing larger by the minute. In this moment the moon was neither the forbidding hole in the stars seen by Bill Anders, nor the ghostly earthlit sphere that had greeted Armstrong, Aldrin, and Collins. He remembered when, on Apollo 10, he and his crewmates were leaving the moon, climbing away so fast that one of them had said, "If we saw this coming in, we'd have to close our eyes." Now he was seeing it. They were making a dive-bombing run on the moon. Shading his eyes from the sun's glare, Cernan could see huge craters, some of which he recognized from three years before. The monocular brought him closer: He could see boulders on the crater walls, ridges curving over the bright horizon. He tried to convey to Evans and Schmitt how unusual it was to see this view. As they drew closer it grew to fantastic dimension, then disappeared into blackness as Apollo 17 headed for lunar orbit. ☾

Cernan would not forget these sights, and he would talk about them for the rest of his life. In his core, he believed it was important to come home with more than just rocks. Now, loading the Rover with supplies, Cernan looked up from his work and saw the earth suspended above the South Massif. No one had ever seen it like this, so close to the horizon; that was because

Taurus-Littrow was so far from the geographic center of the near side. On the way out to the moon, the sight of earth, like a gemstone suspended in dark water, had held him in awe; it was that sight that brought home the fact that there was nothing at all routine about going to the moon for a second time, the single experience that had made him say to himself, *My God, I'm out here again.* Just as he had on Apollo 10, he'd watched it turn on an invisible axis. And once again, he had consciously asked himself whether he understood what he was seeing. His overwhelming feeling was of a sense of purpose, a gut-level awareness that it was simply too beautiful to have happened by accident. He felt as if he were seeing earth as it had appeared in the moment before the Creation, in the mind's eye of God. Now that he was standing on the moon, immersed in the abyss of space—how could the sun shine on him so brightly and not dispel that darkness?—he felt as though he were looking out from within a dream, and the earth was his link with reality. To Cernan it was the most precious possession a man could hold in his memory. He had to say something. ☾

"Oh, man—Hey, Jack, just stop. You owe yourself thirty seconds to look up over the South Massif and look at the earth."

"What? The *earth?!*"

"Just look up there."

"Aaah! You seen one earth, you've seen them all."

It was just like Jack Schmitt to answer his friend's display of emotion with feigned disgust. But in truth, Schmitt had been looking at the earth for three days. Planning to indulge his long-time fascination with meteorology, he'd arranged before the flight to be briefed by air force weather forecasters; he'd launched with satellite photos in the pocket of his space suit. During the trip out, from the moment he floated out of his sleeping bag, he was broadcasting his observations like a human weather satellite. His descriptions filled pages in the air-to-ground transcripts: "That tropical depression I saw earlier north of Borneo is now even more strongly developed, at the tail end of the front that stretches up toward Japan. It really looks like a humdinger from here. . . ." The last thing he needed to look at right now was the earth. He was here to focus on the moon, not because that was his mission, but because it was his profession. And he was impatient to get on with it.

On earth, Cernan and Schmitt had almost spent less time practicing their geologic activities than the purely mechanical tasks that would punctuate them. Paradoxically, that was for the benefit of the geology work: it didn't make sense, for example, to get on and off the Rover at half the pace that they

Schmitt lays out one of the antenna wires for the Surface Electrical Properties experiment, designed to reveal structural information on the material just beneath the surface. The South Massif is at left, Family Mountain at right.

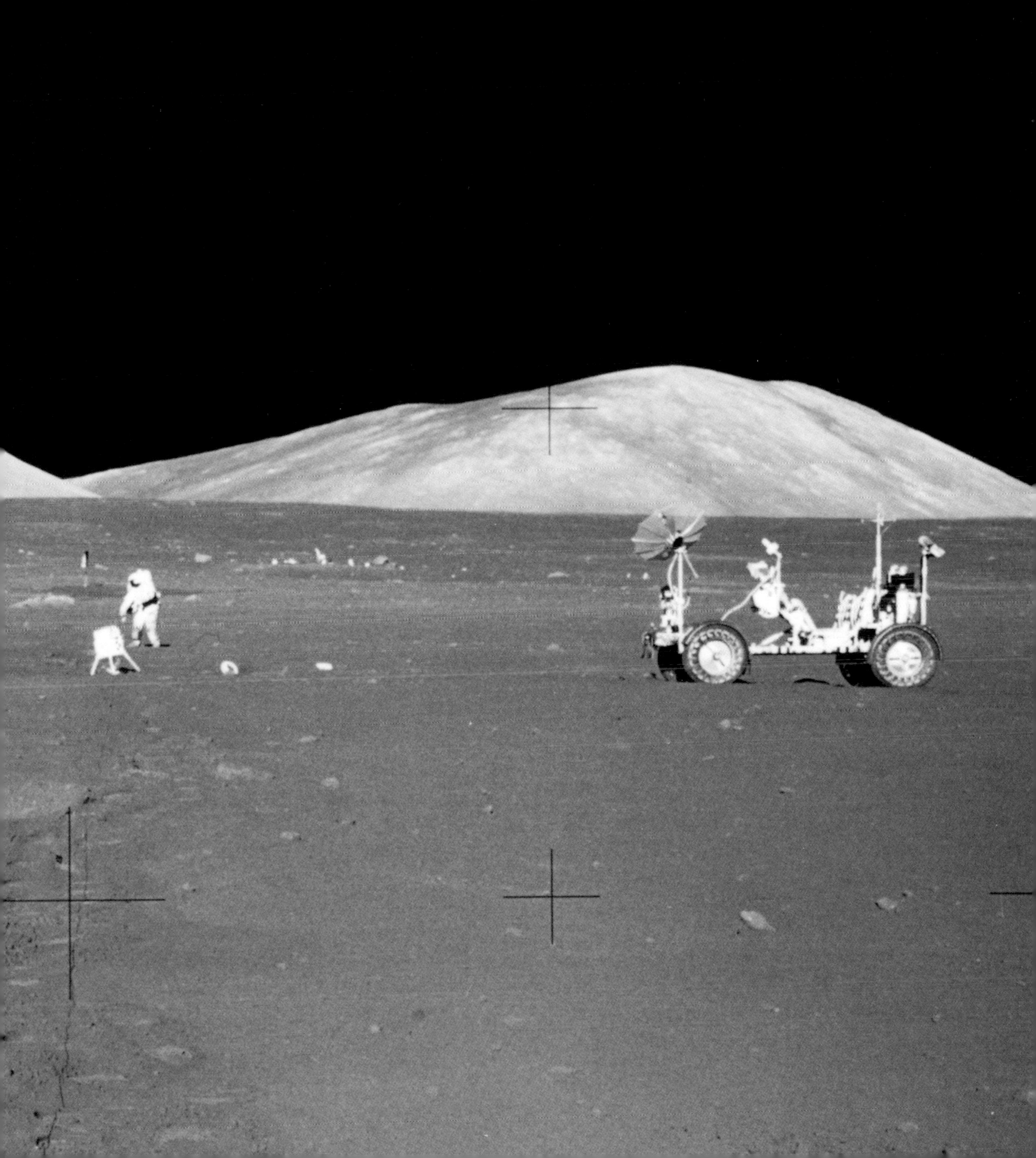

Using a hand-operated jack, Cernan struggles to wrest a deep core sample from the moon's grip *(above, far left)*, as televised by the Rover's camera. Offering help, Schmitt, on the left in subsequent frames, throws himself into the task but ends up tumbling into the dust.

could achieve with practice. By the same token, the men spent hour upon hour laying out their ALSEP experiments, so that they would take up as little of the first moonwalk as possible. But on the moon, predictably, those plans went awry. Fortunately, the heat flow experiment was set up without difficulty —giving Mark Langseth his last chance to confirm the puzzling readings at Hadley. But Schmitt was frustrated by another experiment, the Lunar Gravimeter, designed to show the existence of gravity waves but which now refused to operate. And then there was the lunar drill. Like Dave Scott before him, Gene Cernan was forced into a long, hard battle to obtain a deep core sample. Even the special long-handled jack designed to help Cernan extract the cores seemed to do no good. In mission control, flight director Gerry Griffin watched as Cernan, on his knees, wrestled with the jack. With Cernan's heart rate pushing 145, one controller warned Griffin, "He's eating into that oxygen," and urged that Schmitt help him, "or we're really going to be in trouble." Cernan was in danger of depleting the oxygen reserves designated for the first geology traverse. Schmitt went over to help pump the jack handle, launching his suited body into space and coming down horizontally on the handle, again and again, interrupted by a tumble that sent himself and pieces of stray equipment spinning into the dust—while Schmitt laughed. The flight controllers' concern aside, this was the first sign that something was different about Apollo 17. Both men, and especially Schmitt, took to their work with a kind of physical aggressiveness, even arrogance, rarely seen in any moonwalkers before them, and certainly not on the first day outside; they seemed to have no wariness about the risk of a damaged space suit. It wasn't recklessness—both men had a healthy respect for their situation—but the kind of confidence that comes with familiarity, as if they had done all of this many times before. And in a sense, they had, because five teams of astronauts had gone before them. Once again, the phenomenon of instantaneous evolution that had marked each Apollo mission, not only among the astro-

nauts but the flight controllers and engineers, was at work: Cernan and Schmitt had absorbed the experiences of their predecessors, and now, on the moon, they were building on them.

Finally the moon's grip loosened and the men had their core. Much to Schmitt's frustration, they were an hour behind the timeline, and Cernan's oxygen consumption had forced Gerry Griffin to cut the upcoming geology traverse short; they would go only half as far from *Challenger* as planned. Both men had tired hands and sore fingers, but they showed no signs of slowing down. The Rover's TV camera caught Schmitt as he bounded back to the lander in crazy cartoon leaps, and once again the airwaves filled with song:

"*I was strolling on the moon one day—*"

Cernan joined in, "*In the merry, merry month of—*"

"*December—*"

"No, May," said Cernan. "May is the year of the month."

Schmitt laughed. "That's right, May. *When much to my surprise, a pair of bonny eyes . . .*"

"Sorry guys," said Capcom Bob Parker, unable to resist, "but today *may* be December." Corn on the moon. Jack Schmitt and Gene Cernan weren't letting anything get to them now; they were on an absolute high.

"*Da dada dada dada* deee *da dee . . .*"

TUESDAY, DECEMBER 12
1:12 A.M., HOUSTON TIME
5 DAYS, 4 HOURS,
19 MINUTES MISSION
ELAPSED TIME
INSIDE *CHALLENGER*

The pungent odor of spent gunpowder—that is, the smell of moon dust—filled the cabin as Cernan and Schmitt took off their helmets and gloves. Schmitt knew that on earth, particles of dust are covered with a thin layer of air, but the dust in *Challenger*'s cabin had never been exposed to oxygen. Each grain was still chemically active, just as gunpowder is immediately after

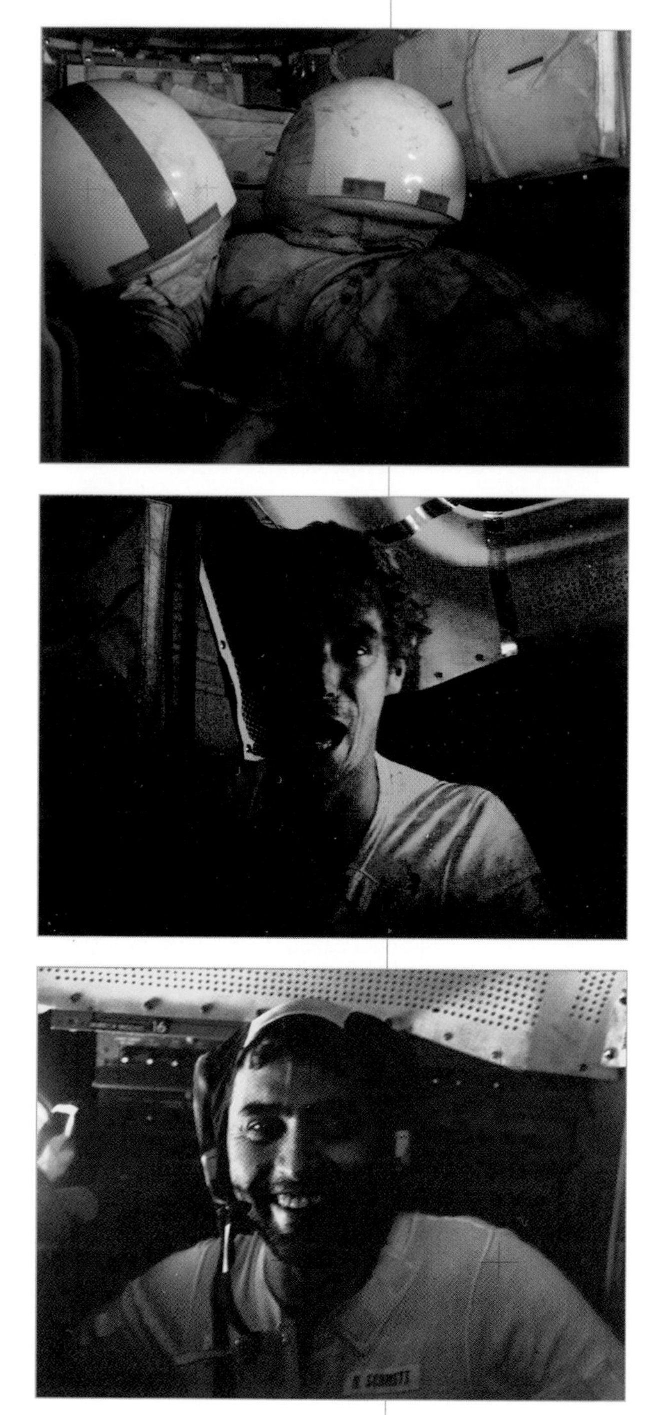

Helmets and space suits stowed in the back of the cabin *(top)*, Cernan *(middle)* and Schmitt prepare for dinner and a night's rest. Cernan's face is smudged with lunar dust, while Schmitt's bears an in-flight beard.

it has been set off, and it reacted with the nasal passages in the same way. Now, to his surprise, he was hit with an attack of hay fever.

It had been a long, hard day. Before the geology traverse, Cernan had accidentally caught his hammer on the Rover's right rear fender and before he realized it, most of the fender was gone. As a result, the entire geology traverse was accompanied by a spray of dust that shot skyward and rained down on the two men. They were so filthy that they had spent fifteen minutes at the bottom of the ladder trying to dust each other off.

Out of their suits now, the men ate dinner; even though the meal was cold (because the LM had no hot water) it was welcome. It had turned out to be a tougher day than either man had expected. Their hands and forearms ached, their fingers were bruised. Schmitt saw blood under Cernan's nails; no doubt partly due to working with the drill. None of this had dampened their spirits out on the surface; excitement had pushed all that discomfort into the background. Schmitt's only worry was whether he would be able to accomplish what he had come for. Dick Gordon had once offered him a bit of wisdom regarding spaceflight: "Time is relentless." And time had reined in him and Cernan today. Their single geology stop, at a medium-sized, block-strewn crater, had lasted only half an hour, just long enough to chip off pieces from a couple of boulders, and to drag the rake sampler through the hard ground for a bagful of cobbles and dust. He hadn't had time to do much observing; mostly he had used stolen moments during other work, and during the traverses, relaying data from his seat on the Rover as they bounced across the dark plains. He had hoped to shed some light on the origin of the dark mantle, but that had eluded him. He still did not know whether it was made of volcanic particles, as the mappers had suspected. Out on the surface, he'd begun to suspect that the mappers had been wrong, that the dark soil might simply be ground-up material from the boulders in the area, which were fairly rich in dark minerals. But the more he looked, the less sure he was—the boulders just weren't dark enough, and if you ground them up, the dust would be even lighter. He hoped there were some experts on earth who were puzzling through these questions in their laboratories. He would have been happy to talk through dinner about it, but mission control had already relayed the questions from the back room and seemed to be letting him and Cernan alone.

The moon was throwing him some curves—of course, that was what his profession was all about. The surprises made it fun. But Schmitt realized that he was in the scientifically vulnerable position of coming all the way to the moon and not learning very much. If something happened and the mission had to be cut short, he and Cernan would leave with a few pieces of basalt—whatever the dark mantle was, those rocks proved there were lava flows underneath it—along with a few bags of soil, a deep core, and a bit of data from a portable geophysics experiment designed to probe the structure of the rock layers beneath the dusty plain. But that wasn't very much, not compared to the secrets the valley of Taurus-Littrow still held.

II: APOLLO AT THE LIMIT

Nothing was ordinary at Caltech, not even a practical joke. Tradition dictates that one morning each spring the graduating seniors disappear, leaving their doors locked with high-tech security measures as a challenge to the ingenuity of the underclassmen. On Senior Ditch Day, as it's called, the object is to break into each fortified room, and once inside—if a student is clever enough to get in—he can do whatever he wants. Over the years this has resulted in some diabolical pranks. A senior returned to his room to find that his furniture had been nailed to the ceiling. Or there was no room at all; the door had been plastered over, painted, and a light fixture installed. Or he opened his door to find a 7-foot weather balloon, full of water. This was not a simple problem. He couldn't move it. He definitely did not want to break it. What did the victim do? He broke it; there was a big flood. Jack Schmitt had witnessed many a Senior Ditch Day prank, and so had several people in mission control. They decided to awaken their slumbering colleague on the moon with a musical reminder of another hallowed Caltech ritual. Every morning during final exams, at 7 A.M., the students in each dorm would tie their stereo systems together, aim the speakers into the courtyard, and play Wagner's "Ride of the Valkyries" full blast. Anyone within a mile radius levitated out of his bed. Inside *Challenger,* the effect wasn't quite as stirring when Schmitt heard the familiar cadence of horns in his earphones, but it did wake him up.

Using geology maps, tape, and clamps—materials available to the astronauts on the moon—engineer Ron Blevins *(right)* helped devise a fix for the Rover's fender, which Cernan broke when he accidentally snagged it with his geology hammer. With Blevins, from left, are John Young, Charlie Duke, Deke Slayton, and Apollo program director Rocco Petrone.

And he was glad to be awake. On earth, Schmitt always needed at least seven hours of sleep (unlike Cernan, who seemed to be able to get by on far less, putting in a long day of training even if he had partied the night before; Schmitt assumed the ability to survive on little sleep was an inborn trait of navy pilots who survived tours of duty on aircraft carriers). Now, even though he had only gotten six hours of intermittent sleep, Schmitt felt rested; thankfully, the soreness in his forearms had disappeared overnight. He would later speculate that the cardiovascular system was so much more efficient in one-sixth g that it cleansed the muscles of lactic acid and other waste products before they could cause any damage. But what lifted Schmitt's spirits most was simply that he and Cernan were still on the moon, that nothing had gone wrong to end the mission, that now, they would finally get a chance to cxplorc thc vallcy of Taurus-Littrow in carncst. Schmitt was actually more excited today, as he climbed down *Challenger*'s ladder, than he had been yesterday, walking on the moon for the first time. After repairing the Rover's broken fender with some maps, gray tape and clamps—a fix devised overnight in one of mission control's back rooms—he and Cernan headed for the slopes of the 7,500-foot South Massif.

There are quite a few places on earth where you can stand in one spot and see more than a mile of vertical relief; the Himalayas, and some parts of the Andes are among them. These mountains testify to the grandest forces on earth, the slow and steady action of plate tectonics. Since 1967 geologists had come to realize that the earth's rigid outer layer is divided into a set of interlocking plates that move along on the fluid upper mantle at a speed of a few centimeters per year, about the rate that your fingernails grow. As these plates collide, move apart, and slide past one another, the continents migrate and the ocean floors are recycled. In the process, most of the earth's major landforms are created. The crust beneath the Pacific Ocean is diving under the west coast of South America, causing massive eruptions at the surface and creating a chain of lofty volcanoes. India is crashing into Asia, pushing up the world's tallest

The astronauts constructed this replacement fender inside the LM and installed it on the Rover. The repair succeeded in keeping the vehicle's right rear wheel from showering moon dust on the men and their equipment as they explored.

mountain range in the process. But the moon has never had continents or plate tectonics, and the geologists who pondered the origin of the Taurus Mountains looked to the titanic forces unleashed by the collision of the Serenitatis asteroid with the moon. But the details of the cataclysm were still entirely beyond their grasp. Had the Massifs simply been thrust upward as unbroken pieces of crust, something like the great limestone blocks of the Himalayas? That had been one working hypothesis for the Apennines, and when Scott and Irwin saw what looked like hundreds of thin layers in the side of Mount Hadley, it had seemed it might be the right one. But since then there had been some doubt about whether those layers were real or simply an optical illusion caused at certain angles of illumination. Another hypothesis was that the Apennines and the Massifs were simply piles of ejecta from their respective impact basins. However they had formed, their majesty assured

that Taurus-Littrow would be remembered as one of the most spectacular places ever visited by human beings. Because the Serenitatis basin had formed before the Imbrium impact, samples from the Taurus Mountains—beginning with the South Massif today, and continuing with the North Massif tomorrow—were the highest priority of the mission. For Cernan and Schmitt, getting to the South Massif meant driving more than 5 miles, the longest traverse ever made on the moon. An hour was budgeted for the journey; never had so much of a moonwalk been allocated to simply getting from one place to another. Seen from the lander, the South Massif was a seemingly bland face of white-gray stone, but yesterday Schmitt had spotted outcrops of different rock types. No one doubted the trip would be worth it.

TUESDAY, DECEMBER 12
6:24 P.M., HOUSTON TIME
5 DAYS, 21 HOURS,
31 MINUTES MISSION
ELAPSED TIME

The ride to the South Massif would stand out in Gene Cernan's mind as one of the most exciting times of the mission. Not because of the geology; he didn't have much time to look at that because he was too busy trying to keep from driving into craters. Anxious to save time, he went as fast as he dared, speeding over blind hill crests without slowing down, ready to swerve if a pothole or rock came into view. For now, Cernan left the geological descriptions to Schmitt. It was remarkable, Cernan thought, how much his lunar module pilot could see from the Rover as it waltzed along, and how much he brought to bear on his observations. Schmitt was so familiar with samples from previous missions that he actually knew them by number.

7:37 P.M.

Seventy-three minutes after leaving the vicinity of the LM, having driven over the suspected landslide and onto a broad scarp, Cernan and Schmitt topped a last rise and descended into a broad trough at the base of the South Massif. Here, a crescent-shaped depression had been christened by Schmitt for the Norwegian Arctic explorer, and later statesman, Fridtjof Nansen. Now it would serve as the scene for an hour-long geologic assault. Had Schmitt and Cernan been able to look back onto the plains, they would have had a difficult time seeing their lunar module. But here in this broad, smooth hollow, *Challenger* and the rest of the valley were hidden from view; this was a place apart from the valley. Perched on this bright slope, Cernan and Schmitt stood at the end of Apollo's longest reach, for Nansen was about as far as two Rover-riding astronauts could travel and still have enough time for a meaningful exploration. It would have surprised no one to learn that Jack Schmitt had always felt the planners were too conservative when they calculated walkback limits, and that before the flight he had pushed, unsuccessfully, to have them relaxed. In any case, the thought of having to walk back to the LM was the furthest thing from his mind. If anything, he felt less at risk here than he had in 1957 and 1958, when he spent eighteen months in the fjords of western Norway working on his doctoral research. Working alone in difficult terrain is something most geologists consider downright foolhardly, but they do it anyway, especially when they are short on time, or have no money to pay a field assistant. In Norway, operating on a shoestring budget, Schmitt climbed alone, often making strenuous treks among sheer rock walls. He worked long into the Arctic summer evening, taking advantage of the midnight sun, because he wanted to get his money's worth out of the effort it took to get there. But Schmitt knew his own limits, and he saved the really difficult climbs until one of his colleagues from the States came to visit; the unsuspecting visitor found himself dragged off on a grueling ascent into the hill country. On the moon, Schmitt knew, he was far from alone, not just because Cernan was here, but because mission control was watching them like a hawk. If either man so much as adjusted the amount of cooling water from his backpack, Houston knew about it. And in his earphones, they heard the steady, even baritone of one Robert Allan Ridley Parker, astronomer-astronaut, Caltech Ph.D., and mission scientist for Apollo 17. Parker and Schmitt were more than a little alike. When he wanted to, the quiet, dark-haired astronomer could serve up the same blunt, needling verbiage Schmitt was famous for. The two scientists maintained a kind of disrespectful rapport, based on mutual insults, that is the coin of the realm at Caltech. The underlying message was, "God, you're incompetent; they really sank the standards when they let you out with a degree." Over the air, Parker maintained a virtual deadpan, but that didn't stop Schmitt from responding as usual. When Parker suggested he take some telephoto pictures of the Massifs "if they look interesting," Schmitt set him straight: "If they look interesting? If they look *interesting?!* Now, what kind of a thing is that to say?" Schmitt knew his friend was in no position to respond in kind, not with the world listening, and that, of course, gave Schmitt's barbs added bite. But in truth, Parker had not been silenced, because he'd inserted little personal touches into Cernan and Schmitt's cuff checklists. For example, at the conclusion of each section there was a message that read, "This is the end, not the beginning." ☾

Nansen crater reminded Schmitt of an alpine valley above the timberline; the craters were smoothed by a blanket of dust that sparkled and threw back the sunlight like new snow. Resting on the surface were hundreds of blocks and fragments which the geologists suspected the landslide had brought down from the heights. He and Cernan had known they would find boulders here, because they had seen them on the orbital photographs. But what was their geologic context? Schmitt cast his gaze on the Massif's whitish face. On the high slope, perhaps a full mile above him, Schmitt could see a few thin, horizontal bands of color—bluish gray and tan-gray, the same colors as the rocks at his feet. There was more here than he and Cernan could possibly investigate in the fifty-odd minutes allotted; they would have to work fast. It was time, rather than the uniqueness of the lunar landscape, that made this fieldwork unlike any Schmitt had ever done. In Norway, he knew that if he had to, he'd scrape together the money to come back for the next field season. Here, the time pressure was monstrous; the drive for total efficiency ruled. Everything he did was for keeps.

8:07 P.M.

At last the mountains of the moon had yielded a rich harvest. Cernan and Schmitt had none of the frustrations that had plagued Scott and Irwin as they searched for boulders on Hadley Delta, or Young and Duke's hunt for the Descartes at Stone Mountain. The time allotted for Nansen was dwindling, raising the inevitable dilemma of whether to exploit a known harvest at the expense of a later, unknown one. That decision was left to the two men on the moon. Cernan decided to accept the back room's offer to take 10 minutes out of a later stop. When Cernan and Schmitt finally headed back to the Rover they had spent 63 minutes at Nansen—nothing compared to a terrestrial field stop, but one of the longest ever made on the moon. For their labors, the men had made an impressive haul; they would leave the South Massif with an entire saddle-bag full of rocks and soil, including a small white fragment, spotted by Schmitt in the middle of a gray boulder, that would prove to be nearly 4.5 billion years old, one of the oldest rocks ever brought back from the moon.

As the Rover pitched and rolled its way back down the mountain—a trip that would be just as memorable as the ride out—Schmitt told Parker he felt confident that the blue-gray and tan-gray rocks he and Cernan had collected could, by extrapolation, be linked to the two layers he'd spotted high on the

The Rover perches on a slope near Nansen crater at the base of the South Massif. The steepness of the mountainside made collecting samples at Nansen strenuous, though rewarding. The astronauts found there a rock as old as the Genesis Rock collected by the Apollo 15 crew in the moon's Apennine mountains.

Massif. Parker responded, "I'm reminded that extrapolation is the nature of our art." Hearing this admission of imprecision from a scientist, Gene Cernan uttered a slow, forced laugh.

9:20 P.M.
SCIENCE OPERATIONS ROOM
MANNED SPACECRAFT CENTER

In the geology back room, Lee Silver studied a television monitor. The pictures from the moon had gotten progressively better with each mission, and the camera's controller, Ed Fendell, had honed his art to the point where he—and anyone else who was watching—could follow Cernan and Schmitt along in their work. Silver could not deny that for himself, Schmitt's presence on the moon gave Apollo 17 a special intensity. He had known Jack Schmitt's father and had met the boy when he was eleven years old. Almost a decade later, he'd lectured to Schmitt at Caltech. Now Schmitt, no longer Silver's student but his colleague, was at work on the surface of another planet.

With two J-missions behind him, Silver was, in his own words, the "old man" of this operation; for Apollo 17 he became the first and only scientist to serve not only on Bill Muehlberger's surface geology team but on the overnight planning team as well. He had not slept since Cernan and Schmitt landed, nor would he until they had left Taurus-Littrow. But Silver had been under pressure long before now. In the three years since he'd become involved with training Apollo astronauts and planning their explorations, the pace of events had been dizzying. The geology team had been designing traverses for Apollo 17 before they had even been able to analyze the rocks from 15 and 16. Silver had seen this phenomenon before; this was classic government science. A federal agency undertakes a massive scientific effort, but is forced by budgetary constraints to press on to new discoveries before the previous ones have been understood. Tied to NASA's mission schedule, the geology team did everything they could to keep up, but Silver, at least, hadn't been completely successful. Some of his Caltech colleagues had developed techniques so sensitive for probing lunar rocks that they could detect individual atoms, but Silver had not had the time or the resources to take advantage of them. And for all the scientists, the slow, painstaking analysis of the samples had lagged far behind the pace of lunar exploration. Some of those samples, Silver knew, contained information that would have affected the planning for each new mission, had there been time to learn of it. For example, no one had really analyzed the peculiar green rocks from Spur crater, which had been on earth for sixteen months. And when this mission was

IN 1972

The first women FBI agents are sworn in.

Bobby Fischer becomes the first American world chess champion, beating Boris Spassky of the U.S.S.R.

Alex Comfort publishes *The Joy of Sex.*

Richard Nixon orders resumption of full-scale bombing of North Vietnam after Paris peace talks reach an impasse.

***Washington Post* reporters Carl Bernstein and Bob Woodward break the Watergate burglary story.**

over the great enterprise called Apollo was shutting down, with so much of the moon left unexplored—and that was also classic government science.

If the exploration of the moon had put a professional strain on him and his colleagues, Silver understood that it had also exacted a high cost in personal terms, in strained marriages and disrupted families, for people throughout the program. Silver himself had been in a period of great turmoil. His responsibilities at Caltech were crushing; his personal life was rocked by crisis. One night at around 3 A.M. in front of the King's Inn on NASA Road One, Silver told Schmitt he couldn't keep this up, that he was pulling out of the training program. Schmitt became very upset, and Silver suddenly realized how much the man was depending on him. He decided to stay on, though he was never able to have the same involvement he'd had on Apollo 15.

That was the only time in the past sixteen months that Silver glimpsed a crack in Schmitt's stubborn armor. Schmitt himself would never admit he felt pressure, but there had been some subtle indications. Silver had seen Schmitt's demeanor change when he made it onto the 17 crew. There was less joking around than when he had been a backup crewman; this was serious

As far as Silver was concerned, Schmitt was doing a beautiful job. And Cernan was holding his own.

business. And Silver knew about the pressure from the scientific community to get him on the flight; he also knew that Schmitt had won his seat at the expense of a very popular pilot-astronaut. Silver could not guess at how the other astronauts viewed the decision, but he did sense the scrutiny of the scientific community, who were watching Schmitt and waiting to see whether he would live up to his billing. Some of them, Silver was sure, were probably half-hoping he would screw up.

As far as Silver was concerned, Schmitt was doing a beautiful job. And Cernan was holding his own. Early on, some of the geologists had privately worried that Cernan, as a mission commander who enjoyed the limelight, might be reluctant to yield center stage to his lunar module pilot. They had underestimated Cernan's flexibility. Cernan was so accepting of the situation

As televised by the Rover, Schmitt struggles to pick up a bag of samples that he has dropped at a crater named Lara. By this time, well into the mission's second moonwalk, Schmitt's hands had become tired from the strain of opening and closing them inside his pressurized spacesuit gloves.

that Silver had felt it necessary to take him aside and encourage him not to let Schmitt be "his geologist," in effect, but to be more of a partner—which, having seen Cernan's powers of observation, Silver knew he could be. Watching the pair at work on the moon, Silver was impressed by how Cernan complemented Schmitt's focus on detail with a view of the big picture. But if, at times, Cernan almost seemed to be functioning as a field assistant—"Get your hammer," Schmitt told him at one point, "we're going to need it"—that was another measure of this commander's flexibility. The fact was that all through the moonwalks, Cernan had been making first-rate observations, calling Schmitt's attention to intricate fracture patterns in the rocks, changes in texture, subtle hints of bedding. They had become a well-honed field geology team.

Just now, Cernan and Schmitt were back on the valley floor, near a 1,500-foot-diameter crater named for *Doctor Zhivago*'s Lara. And Schmitt was doing what he did at every stop—looking around, sizing up what he saw, and relaying the information to the back room so that they could plan their goals accordingly.

"Bob, I'm at the east-southeast rim of a thirty-meter crater, in the light mantle, up on the scarp and maybe two hundred meters from the rim of Lara. . . . There's only about a half-centimeter of gray cover over very white material that forms the rim. . . ."

This was a new way of doing business, made possible by the fact that for the first time, there was a trained scientist on the moon. But it wasn't long before Schmitt began to lose the upper hand in his a *mano a mano* with the moon. There was too little time for detailed, systematic samples, but while Cernan hammered in a 32-inch-long core tube, Schmitt made his own solo investigations. He made a valiant effort to collect rocks by himself, holding his long-handled scoop in one hand and a Teflon baggie in the other. Every time he tried to set the scoop down, it fell over. Then he knocked over a big

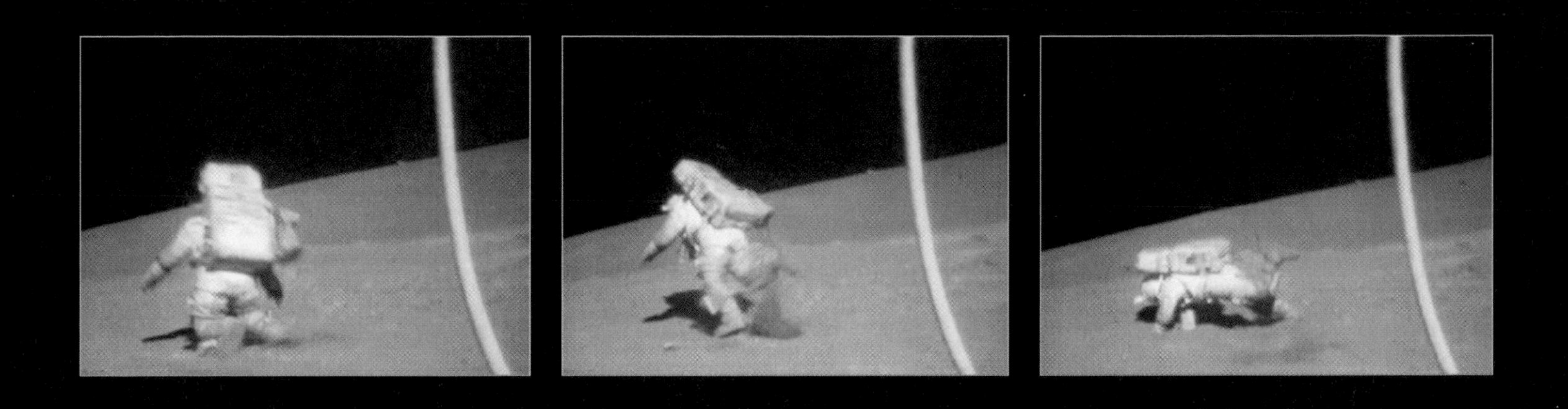

collection bag, scattering samples on the ground. Schmitt went down on hands and knees, rounded up the samples, and put them back in the bag. Then he leaned back on his heels to get his massive backpack over his feet, and launched himself back to a standing position—and dropped the collection bag in the process. Schmitt got down on one knee to retrieve the collection bag, but he stumbled and fell on his chest. At that point Bob Parker started calling Schmitt "Twinkletoes." Later, he radioed that the switchboard at the Manned Spacecraft Center was lighting up with calls from the Houston Ballet Foundation requesting Schmitt's services for the next season. Seizing the opportunity to audition, Schmitt did a graceful leap—and plummeted to the dust, laughing.

But that was a temporary setback. Minutes later, as Cernan and Schmitt headed for their next stop at a crater called Shorty, Silver and the rest of the back room were reminded how good their surrogates at Taurus-Littrow really were. At several points along the traverse, Cernan had stopped driving just long enough to allow Schmitt, still seated, to reach down with a special scoop, which had small bags attached, and collect a sample without getting off the Rover. One of these "Rover samples" had been planned while the men drove across the outer portion of the suspected landslide, which the geologists called by the noncommittal name "light mantle." But at the last minute, Muehlberger's team had sent word to cancel it in favor of more time at Shorty. Schmitt and Cernan decided to stop anyway—and it was a fortunate decision. That scoop of light-colored soil would realize one of the geologists' most cherished hopes for Apollo 17. All along, the geologists had suspected that the landslide—this sample helped prove that theory—had been triggered sometime in the distant past, by flying debris—ejecta from a distant impact—striking the South Massif. And the impact, they believed, was probably Tycho, 1,300 miles to the southwest. Analyzing the landslide material, the scientists would determine it had been lying on the valley floor for 109 million years, affording a compelling, if not airtight, time for Tycho's formation. In a stroke of incredible good luck, the trip to Taurus-Littrow appeared to have given the geologists one of the most important dates in lunar history. By exercising their own judgment as trained lunar field geologists, Cernan and Schmitt had beautifully shown their value in the exploration. And this moonwalk wasn't over.

10:18 P.M.

"Okay, Houston. Shorty is clearly a darker-rimmed crater. The inner wall is quite blocky. . . . And the impression I have of the mounds in the bottom is that they look like slump masses that may have come off the side." Jack Schmitt, about to explore the 360-foot crater, made his report to the back room. On the orbital photos, Shorty was ringed by a dark halo that stood out

against the light mantle. The geologists suspected it was a volcanic explosion crater, perhaps one of the sources of the dark-mantle deposits. They knew it might well turn out to be just another impact crater, and that the halo was simply an ejecta blanket of dark soil. But the possibility that Shorty might at last reveal evidence of recent volcanism was simply too exciting to ignore.

As entertaining as Jack Schmitt's comedy of errors at Lara may have been for those watching on earth, it had greatly frustrated him. In part, he'd gotten into trouble because he'd had to work alone; he'd practiced a little solo sampling before the flight, but not nearly enough to become efficient. And also, he was getting tired. His hands and arms ached too much for him to keep his grip on things. On the drive to Shorty, he had told himself he would forgo any more attempts at solo work for a while. Now that he was here, his spirits were lifted by the sight of Shorty itself. The crater was big and deep, its walls littered with rocks, its floor a maze of rubble. The lighting was dramatic. But Schmitt knew he and Cernan would have to work fast; having spent

Schmitt and the Rover are dwarfed by Shorty crater, some 360 feet across. It was here, near the crater's rim, that Schmitt's boot uncovered a remarkable layer of orange soil *(page 230)*, conclusive evidence of eruptions from deep within the moon in eons past.

The orange soil that Schmitt discovered shows clearly in this trench he dug to assess how deeply the color extended into the surface. At left is the gnomon, a photographic reference device used to judge scale, color, and slope.

an extra 10 minutes back at Nansen, they would have only half an hour here.

A large, intensely fractured boulder near the crater rim caught Schmitt's eye. Once Cernan was finished with his Rover chores, they would sample it; for now, Schmitt would take a panorama from the rim. He moved closer to the boulder and stopped, preparing to take his pictures. Instinctively, he scanned the area around his feet for details, and suddenly he saw that his boots had scuffed away the dust. Instead of the usual monotonous gray, he saw vivid, bright orange. For a moment he wondered whether it was an illusion. At an earlier stop, he'd seen a patch of color on the surface that turned out to be a reflection off some gold foil on the Rover. He lifted his outer visor partway and looked again. It was *real.* "There is *orange soil!*"

"Don't move until I see it," said Cernan, still working at the Rover.

In the geology back room, Lee Silver *(right)* enthuses about the orange soil to geologist Jim Head, who is watching the action on a TV monitor. Head worked for Bellcomm, a think tank that NASA hired to help plan the astronauts' lunar explorations.

Privately, he wondered whether Schmitt had been out in the sun a little too long—until he turned around. *"Hey, it is!! I can see it from here!!"* Cernan's voice had risen a full octave.

"It's *orange,*" Schmitt said, as if he couldn't believe what he had found. It glowed like a highway sign in the glare of headlights.

Cernan wondered aloud, "How could there be orange soil on the moon?" Then he answered his own question: "It's been *oxidized.*" Yes, Schmitt agreed, it looked like the rusted soil you sometimes see in the desert. But if Cernan was right, what had done it? How, on a world devoid of water and air, could there be rust? The answer was volcanic gases. You can walk around the slopes of the Kilauea volcano in Hawaii, as Cernan and Schmitt had done in training, and see chunks of dark lava that have been turned bright orange, red, or yellow, by hot mineral-laden gases escaping from the depths. At that moment, in the back room, Lee Silver was practically jumping up and down with glee, saying, "That's it! That's the volcanic vent!" And Bill Muehlberger was saying, "Somebody unplug Silver."

Unaware of this ecstatic outburst, Schmitt immediately went to work. First he dug a trench into the orange, to find out the extent of the deposit, and he saw that the orange lay along a ellipse-shaped zone that ran parallel to the crater rim. Perfect; the same thing happens in Hawaii. The fractures that allow the gases to percolate upward often follow the perimeter of the volcanic craters, and as a result, so does the pattern of altered soil. Furthermore, the ellipse was divided into zones of color: The orange merged with yellow toward the edges, while in the center the soil was almost crimson. Such *zoning,* caused by different concentrations of gas, was another hallmark of volcanic action. "Man, if there ever was a—" Schmitt laughed. "I'm not going to say it." Yes, he was, even if he turned out to be wrong. "If I ever saw a classic alteration halo around a volcanic crater, this is it." When he arrived at Shorty, Schmitt had been all but certain that it was an impact crater; now there was a very real possibility that Shorty was volcanic. But unless he and Cernan could get an extension, they wouldn't have time to prove it. At the same moment, Lee Silver was telling Jim Lovell, "We need more time." But there wasn't any. The ever-tightening circle of oxygen consumption and walkback limit was closing in on Cernan and Schmitt. Unlike Nansen, the option for an extension at Shorty simply did not exist. ☾

Cernan, finished with his Rover chores, at last joined Schmitt to collect samples. Schmitt dug his scoop into the orange and found it did not behave like dust; it broke apart in angular fragments. He dropped most of the first scoop before he could pour it into Cernan's sample bag. He told himself to slow down. A second try filled the bag with orange clumps. Then he captured the gray soil at the margins, for comparison. The next step—and perhaps the most important—was to find out how deep the deposit went, and what might be underneath it. The back room was already thinking about that; Parker passed up a request for a core tube. And at the same time, he informed the men that they had 20 minutes left.

Cernan drew his hammer. Schmitt watched as he pounded the core as hard as he could; the tube penetrated less than half an inch with each whack. "I'd offer to hit it," Schmitt said, "but I don't think I can, my hands are so tired." Cernan persisted and drove the core most of the way in.

Cernan drew his hammer. Schmitt watched as he pounded the core as hard as he could.

Thankfully the core came up without a struggle. "Even the core tube is red!" But not all of it. Schmitt was startled to see that the bottom of the tube was coated with dark purple-gray, almost black soil. For a moment, before capping it, the two men cradled the 3-foot core, holding in their hands one of Apollo's most exciting discoveries, as the scientists would realize when it and the other samples were on earth. Their impact would not be diminished one bit when it became clear that Schmitt had been wrong when he thought Shorty was a volcanic vent, that in fact it was a run-of-the-mill impact crater. The orange soil was anything but ordinary: It was made of tiny beads of glass that had once been molten droplets, poor in silica but rich in titanium and iron, propelled high into the lunar sky by a spectacular form of eruption called a fire fountain.

The origin of the fountain was a form of lava that contained dissolved volcanic gases. As it ascended from deep within the moon to the surface, the effect was that of shaking up a bottle of soda and then uncapping it: The gas rapidly came out of solution, propelling molten rock high into the lunar sky. Just as water pressure in a decorative fountain causes the liquid to break up into droplets, this so-called fire fountain was composed of an intensely hot spray. In the weak gravity, the droplets arced hundreds or perhaps thousands of feet through the vacuum. During their flight, they cooled into tiny glass spheres, which rained down on the valley of Taurus-Littrow. Different spheres cooled more slowly than others, affecting their final mineral content. Some of the beads were orange, some were crimson, not from oxidation, but due to a high titanium content. Others—many of them—were richer in iron, partially *devitrified* (that is, crystalline), and colored black. Those beads, which had startled Schmitt when he saw them at the bottom of the core tube, were the answer to the riddle of the dark mantle. The soils of the valley of Taurus-Littrow—and, by implication, the southeastern portion of Mare Serenitatis—were dark because they contained beads of dark, volcanic glass. But the fire fountains had not sprayed forth recently, as the geologists had hoped; the glass beads were 3.5 billion years old. Untold millions of years later, probably, the deposit had been covered over by a fresh lava flow, which protected the strange beads from being mixed with ordinary lunar soil by micrometeorite rain. They remained hidden for eons, until 19 million years ago, when the Shorty impact excavated them and brought them to the surface, where they awaited the footsteps of a geologist from earth. And the orange soil discovery would lead the geologists to analyze the beads of green glass from Apollo 15, which would also turn out to have formed by fire fountains. In both cases, the lava had come from hundreds of miles down in the lunar mantle, far deeper than any other Apollo samples. And on the surfaces of the beads, analysts found coatings of volatile elements—the first such elements found in any lunar samples—which were solid remnants of the original volcanic gases that had powered the eruptions. Knowing that these gases had existed within the moon would cause scientists to revise their models for its evolution. ☾

The fact that Schmitt had been wrong when he suspected Shorty was a volcanic crater did not lessen the impact of his work. Despite the fact that Schmitt felt he had barely enough time to think, he and Cernan had performed so well that Muehlberger's planning team, after deliberating all night on whether to send the men back to Shorty the next day, decided that would not be necessary. The men had done everything needed to decipher Shorty's story. Schmitt even managed, as the remaining minutes ticked away, to gather pieces of the basalt boulder that had caught his eye to begin with. Meanwhile, Cernan snapped a spectacular series of panoramic pictures, and to his surprise, he spotted streaks of orange and dark gray on the crater walls. He was still trying to tell the back room what he saw, talking so fast that he was

A cluster of the lunar module's maneuvering thrusters frame the American flag and the Lunar Rover, parked for the "night" after the conclusion of the second moonwalk. After their sleep period, the astronauts were scheduled to suit up again for their third and final EVA.

almost out of breath, when Parker cut him off: "You can talk about it when you get home."

5:12 A.M.

Cernan and Schmitt were bedding down for a well-deserved rest when Joe Allen, taking the night shift in mission control, told them that the flight controllers were still marveling at the beautiful television pictures coming down from Taurus-Littrow. They were fascinated by the footprints and the Rover tracks, and they were speculating on what might someday disturb them, after Cernan and Schmitt had departed.

"That's an interesting thought, Joe," Cernan replied, "but I think we all know that somewhere, someday, someone will be here to disturb those tracks."

"No doubt about it, Geno," said Allen. They all knew that the young physicist in mission control would never have the chance himself. But Schmitt felt the need to stress that if it was too late for Joe Allen, there was still reason to be hopeful. "Don't be too pessimistic, Joe. I think it will happen."

"Oh, there's no doubt about that," Allen agreed. "But it's fun to think about what sort of device will ultimately disturb your tracks."

"Well," Schmitt said, "that device may look something like your little boy."

Allen laughed, thinking of his four-year-old son, David: "He'd make short work of them."

Cernan was in his hammock now, and Schmitt was about to get into his. They would have their eight hours rest, and then would come the final moonwalk. Schmitt took a last look at the earth through the overhead window, and rolled the shade closed. He told Allen with gentle irony, "Tomorrow we answer all the unanswered questions. Right?"

III: WITNESS TO THE EARTHRISE

WEDNESDAY, DECEMBER 13
EL LAGO, TEXAS

For days now, Jan Evans had ducked outside in hopes of seeing the moon, but instead she saw an impenetrable gray drizzle. Finally, on Wednesday the thirteenth, a cold front moved in and the sky cleared, and there it was, ripening toward full. Finally, she could see where Ron was. His moon wasn't the one she remembered from her Topeka, Kansas, childhood, on those winter nights during World War II when the whole town would turn off its lights during air raid drills. In her living room, her mother had draped a towel over the Philco radio with its big yellow dial. Outside, she could see the air raid warden patrolling the street. The trees were coated with ice, sparkling in the light of a full moon. That would always be her moon, not the one Ron was circling now. She would have loved to be a mouse in the corner, up there with him, if the thought hadn't scared her half to death. And Jan was quite content to listen to the squawk box that brought Ron to her. And she heard a lot of him, because Ron had decided to make a special effort to share his experience with the world as it was happening. Whether the world was listening was an open question; here at home the networks were giving Apollo 17 only spotty coverage at best. In New York, the sixth

For days now, Jan Evans had ducked outside in hopes of seeing the moon, but instead she saw an impenetrable gray drizzle.

lunar landing made for only a brief interruption to the soap opera "Love Is a Many Splendored Thing," and the Johnny Mann Singers drew more viewers than live images of two explorers on the moon. But that was far from Jan's consciousness. Here in El Lago, with a house full of people—and a lawn full of reporters—she was having a ball following her husband's adventures. In between streams of technical jargon, she heard his excited chatter, his humming and whistling. She knew he was very busy, that the flight plan had him running from the moment he awoke until bedtime. On the way out to the moon, Ron had been on watch and had overslept for a solid hour, while mission control tried again and again to wake him up. But that was nothing new. Jan told everyone the story of how in the navy he'd been such a sound sleeper that his bunkmates on the carrier would leave a Baby Ben in a metal wash basin right next to his ear. But Ron wasn't oversleeping now, not with all that work to do. At least he was eating well, and that was a good thing, because if there was a chowhound on this crew, it was her husband. Before Gene and Jack had departed, there had been a small crisis when Ron lost his scissors somewhere in the command module, leaving him without any means of opening food bags for three days until they got

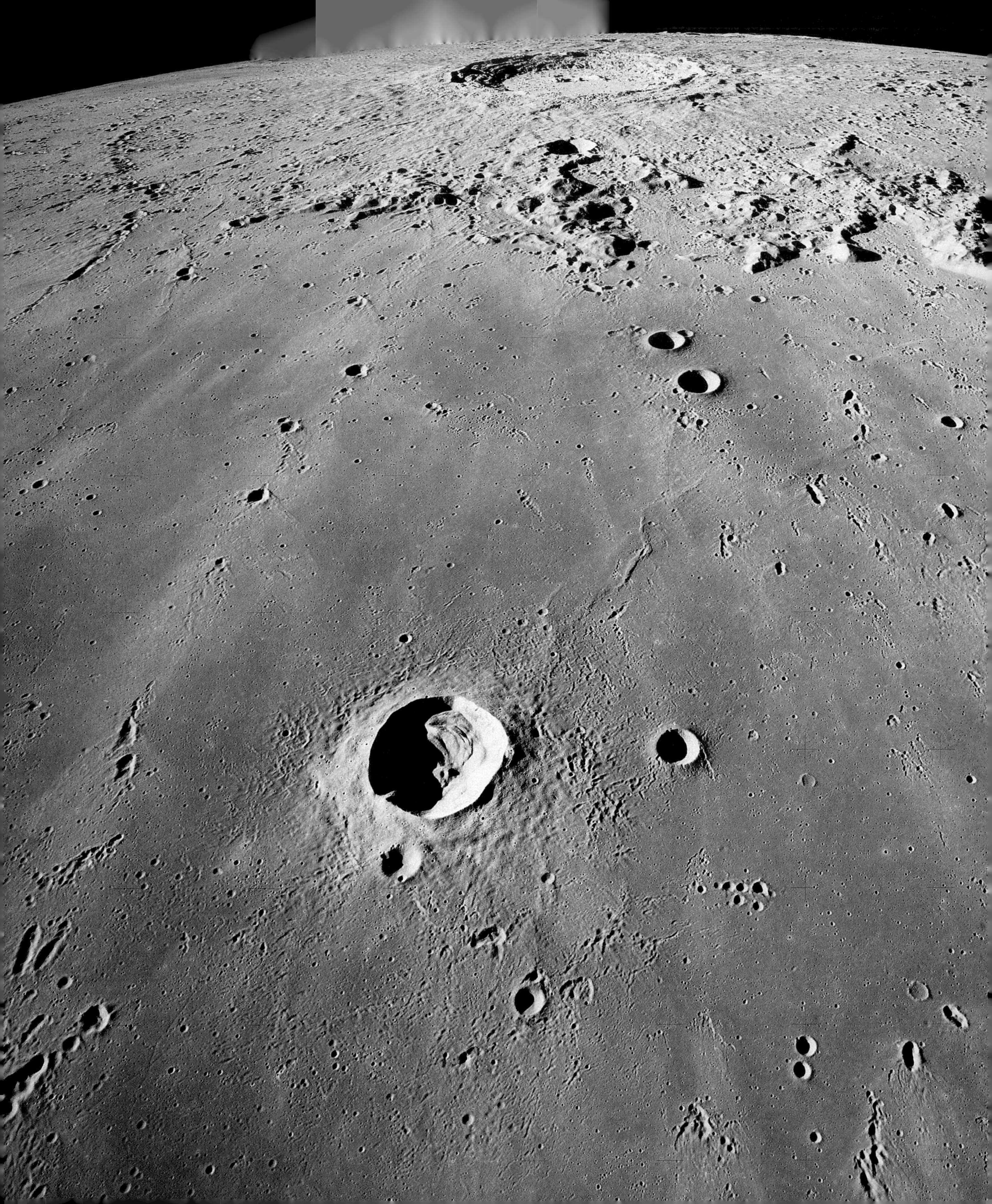

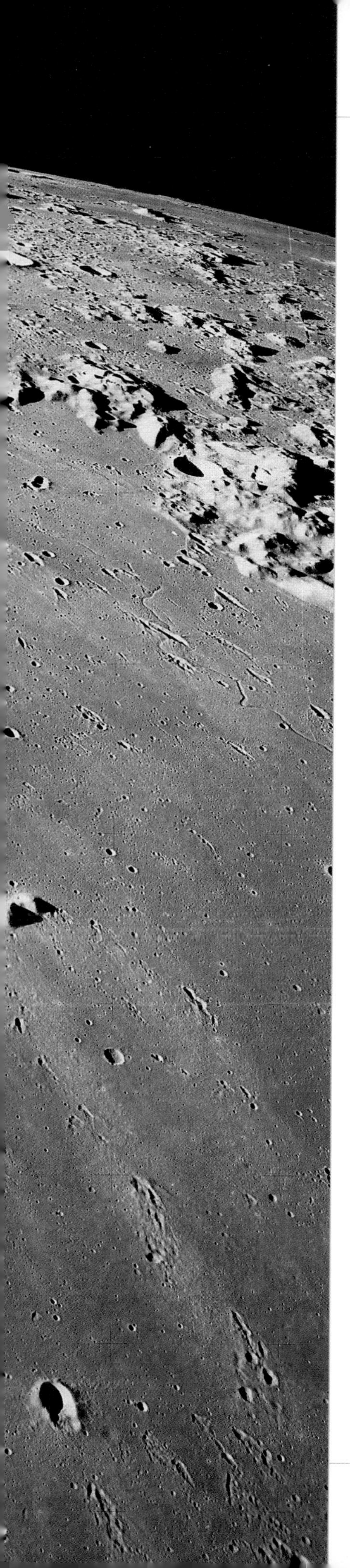

Aiming *America*'s mapping camera southward across the Sea of Rains, Ron Evans photographed the 58-mile-wide Copernicus crater, seen here on the horizon. Bright streaks and clusters of small craters were created by ejecta from the impact that formed Copernicus roughly one billion years ago.

back. They saved him by giving one of their pairs to him. But Jan had a laugh when she learned that on one of his solo days the flight plan hadn't left him any time for lunch.

After the Original 19 were selected, Ron had gone to Houston to look for a house, and she could still remember the day he called up and said, "You have to get an air conditioner for the car." She was afraid all this astronaut business had gone to his head—Air-conditioning for the car? On his salary?—but the next time he called, he was even more insistent: "I'm not kidding, you've got to get air-conditioning!" And when she and the children had trekked from San Diego, and across most of Texas, and arrived in tropical heat, she understood what he meant—air-conditioning was necessary to support human life. Ron was the last person who would let the title of astronaut go to his head.

The astronaut life had not been tough on her, the way it had on other wives. From the beginning, she considered it a luxury to have him home on weekends instead of gone for eight or nine months at a time, the way he had been when he was serving in Vietnam. And the risks of spaceflight were something she rarely thought about. She had seen the way NASA operated, and she trusted the system; she had come to feel that Ron had been in far greater danger flying missions over the jungle than he was now, a quarter-million miles from home. She knew Ron had always felt that being an astronaut he had gotten the best deal out of life; she also knew he had friends who had not been so lucky and were at this moment lying in North Vietnamese prisons. And she knew he had always felt a little guilty about that. But he had wanted to fly in space like nothing else she had ever known him to want. It had kept him going for six years, through the unglamorous support-crew duties of writing checklists, going to meetings, and setting switches before spacecraft tests; through the night shifts in mission control at the Capcom's mike. Now that he was up there, and his voice was coming to her from across that amazing distance, he sounded like the same old Ron she had met in high school—except that he was having the time of his life.

Jan didn't know the flight plan very well, but she knew the critical phases, things that were essential to carrying out the mission. She followed along when she could, and she got up when Ron got up, and went to bed when he did. She was grateful to Deke Slayton, who called every evening to review the day's events with her. "They made a great burn today," he would say. Or, "Things are going really well." Looking back later, she would say that Slayton was nothing less than the glue that held the astronaut families together.

ABOARD THE COMMAND MODULE *AMERICA*, IN LUNAR ORBIT

Most of the time, it didn't even feel like flying, except for once, when he lit up the SPS engine to change the orientation of his orbit, and he felt the sensation of actually flying a space ship; it felt like going full afterburner in a T-38. The rest of the time it was a lovely quiet. His ship was performing flawlessly, and he was proud of it—and what it stood for. He and his crewmates had named their command module *America*, and to Evans it was the perfect choice. The world should know it was the United States up here, that there was something else his country was giving to history besides a war, something uplifting. During training, Cernan had started calling him "Captain America." And if Evans was an unabashed patriot, then he wasn't afraid to say, on the record, that he thought the Nixon administration had paid lip service to the space program without giving any real support, or that the politicians

In the geology back room, Jim Lovell *(left)*, Bill Muehlberger *(center)*, and Gordon Swann *(seated)* await Apollo's final moonwalk on December 13, 1972, eagerly anticipating the astronauts' visit to the slopes of the North Massif.

who opposed the program were being swayed by people who thought the money would be better spent on welfare—and he didn't hesitate to call those people kooks. Maybe that kind of outspokenness, almost unheard of among astronauts, came from having risked his neck for his country flying missions over Vietnam. But back then, he'd been just another unknown aviator.

And if he let his excitement hang out over the airwaves, that was just the way he liked to do things. Why not put aside the image of the cool space traveler? He knew people saw astronauts as reticent; he was changing that. When he wasn't enthusing to Houston, the onboard tape recorder caught him talking up a storm, to himself. "What happened to my grits? I lost my grits. There they are again. Anything you drop up here, it just disappears, flat disappears." He was trying to grow a beard for his son, Jon, but it itched so much he was tempted to shave it off. "Just be comfortable," Ken Mattingly in Houston said. "You got another week to go." It was a long mission. Even now, more than 6 days into it, he was only halfway through. And he was beginning to dislike his own company. "Man, I stink! Whew! Soap doesn't do any good. I'll be glad when the guys get back with the deodorant. Took all the deodorant with them to the surface. Outstanding stuff."

Evans made no secret of the fact that he would have liked to walk on the moon—but there was no doubt he meant it when he said that getting within 9 miles was good enough. He was doing his part, as Farouk El-Baz's last protégé in lunar orbit. And he thought about his crewmates and asked about their progress on the surface. Evans had known Cernan since they were both at the Naval Postgraduate School in Monterey, California. They'd applied for the astronaut program together, and when Cernan made it, they went out and got drunk together. It had been hard for them to lose Joe Engle, but when Schmitt came on the crew, they made room for him. Evans liked Schmitt well enough, and he gave him a lot of credit for hard work and persistence. "Dr. Rock," he called him. And Evans knew that you can take astronauts with the most diverse personalities and put them in the same spacecraft, and they will function as a team because of one thing: The mission comes first. ☾

1:37 P.M., HOUSTON TIME
6 DAYS, 16 HOURS,
44 MINUTES MISSION
ELAPSED TIME

The phases of the moon and earth are complementary. When the moon is new, a full earth shines in the lunar night. As we see our satellite wax through crescent and first quarter to full, our own world wanes in the moon's sky until it is lost in blackness. On Wednesday, December 13, Gene Cernan pulled back the shade on *Challenger*'s rendezvous window and saw a blue-and-white crescent gleaming above the South Massif. He could barely make out a cloud system over the southeastern United States. In mission control, Capcom Gordon Fullerton confirmed that yes, that's what the satellite picture showed. How amazing, Cernan thought, to be standing on the moon, giving his home planet a weather report.

After a quick breakfast, he and Schmitt—by this time, a little weary—got ready for Apollo's final moonwalk. They were an hour late getting started, but there wasn't any particular need to rush.

At the very first mission review meeting Jack Schmitt had lobbied for a fourth moonwalk. He'd worked on the initial design studies for the J-missions, and he knew that four excursions had been part of those original plans, and now he had no qualms about telling the engineers what their hardware was capable of. On paper, he was right; you could squeeze in a fourth moonwalk, but it left you with very tight reserves of water and battery

How amazing, Cernan thought, to be standing on the moon, giving his home planet a weather report.

power in case of an emergency (like a blown water tank). The suits and backpacks hadn't been tested for the extra trip outside, though everyone suspected they would hold up. The planners spent the better part of a day with Schmitt, trying to talk him out of the idea, but to no avail; he knew he was right. Cernan just hung back and let Schmitt say his piece. Finally, Apollo spacecraft program manager Owen Morris asked Cernan for his own recommendation. "I think we can do it," Cernan said, "but I'll respect your judgment." Schmitt shot Cernan a dirty glance; that was the end of it.

And if Schmitt felt some lingering frustration at using Apollo to what he saw as less than its full capability, he no longer feared having to leave the moon without enough information. This day they would have a good seven hours on the surface, enough time to sample the North Massif, and the Sculptured Hills, and a couple of craters on the valley floor. And then, no more. So there they were, in their hard suits, moving like tin soldiers as they prepared to open the hatch. When Cernan emerged from the tiny cabin at around 4:30 P.M., he saw a small sign taped to the landing leg. He did not know who had put it there, but it must have been one of the checkout crew at the Cape. On each moonwalk, he'd paused to read it aloud as a kind of public prayer. Now he read it again: "Godspeed the crew of Apollo 17." ☾

5:37 P.M.

The Rover headed north across a sun-drenched moonscape, past the giant crater Henry, and when it reached Turning Point Rock the men headed east toward their destination, a big, dark boulder perched on the side of the North Massif. They had all known the boulder would be there; it was visible on the Apollo 15 photographs, where it was a tiny dot trailed by a 1,500-foot-long furrow that recorded its plunge down the mountainside. As they got closer Cernan and Schmitt could see that it was actually five separate boulders; it had broken apart before coming to rest near the bottom of the Massif.

As always, the Rover gave them no hint of the severe slope. Cernan, on the uphill side, could barely raise his suited body out of his seat. Schmitt, meanwhile, thought for a moment that he might tumble down the hill. While Cernan struggled through his chores at the Rover, Schmitt climbed a few yards upslope and raised his outer visor, which was now badly scratched by dust, to peer at the wall of stone. It was just like the tan-gray rocks he'd seen yesterday at the South Massif, except that it had giant vesicles—holes where gas bubbles had once been—which meant that it had once been molten. But the holes weren't round, as you would see in a lava flow that cooled in place, but flattened ellipses, meaning that somehow, the rock had flowed before it had had a chance to solidify. Sometime later a great pressure had acted on the rock, for Schmitt could see fractures branching through it. It was difficult for him to tell much more. On earth, geologists must peer at their work through a variety of obscuring materials such as moss or lichen, or, in the desert, a smooth "varnish" of oxides created by the action of wind and rain. On the moon, rocks develop a brownish patina of impact glass from millions of years of micrometeorite rain. To get a good look at this rock, Schmitt knew, Cernan would have to put the hammer to it.

In the back room, Bill Muehlberger's team watched as Schmitt methodically examined the boulder. Every so often he would change his mind about what he saw, verbalizing the thought process that goes on in the mind of every geologist in the field. They knew that Schmitt had attacked complex

With a leap that lands him gently in the right-hand seat, Schmitt mounts the Rover during the third moonwalk. In his hand is a specially designed scoop that enables him to gather samples without dismounting the Rover.

problems like this one in Norway, and in Sudbury, Ontario, where the remains of a giant impact crater are preserved. Years later, Schmitt would say that his edge on the moon was that he was able to react to the millions of bits of visual data and sort out the most significant few, and do it within seconds. The mind, he would say, is not in a space suit.

Now the television camera caught him taking pictures, standing in one of the big hollows of the boulder track. With his golden outer visor raised, his face was clearly visible within the clear inner helmet, framed by his communications hat and darkened by a week-old beard. As he spoke to Cernan, something about a piece of gear, his mouth could be seen forming the words. Suddenly he was no longer a faceless, mirror-helmeted astronaut, but Harrison H. Schmitt, a geologist at the ultimate field site. He noticed the camera pointed at him, looked straight at it, and mugged a smile. Then, having spotted something new to investigate, he turned to his left and headed out of view.

●◐◐○○○◑◑●

Working together now, Cernan and Schmitt attacked the boulder, Cernan wielding the hammer. It was ironic; Schmitt had helped design that hammer, but he couldn't hold it; the handle was too big for him to grip. The men had agreed in training that Cernan, with his massive hands, would be the mission's hammer bearer. The only thing Schmitt had to do was teach Cernan to hit the edge of a rock, which was much easier to chip, instead of pounding away, brute-force style, at the middle.

Schmitt made his way out of the shadow and around to the boulder's sunlit face, using his scoop like a walking stick. The more he looked, the more details he noticed. Within the tan-gray boulder were huge pieces of blue-gray, caught up like pebbles in the tar of a new road. They looked just like the blue-gray rocks they'd picked up at the South Massif. And within the blue-gray there were smaller, white chunks that looked like anorthosite, perhaps pieces of primordial crust. Cernan thought the entire boulder was one enormous breccia, but Schmitt wasn't so sure; the tan-gray rock had definitely been molten, and Schmitt thought it had formed when a pocket of magma caught up pieces of the blue-gray breccia within it. Whatever it was, this boulder had a story to tell, and if they were going to decipher it they would need as much time as they could get. Parker informed them that they could have the lavish allotment of 1 hour and 20 minutes, but Schmitt was reluctant to take it all; it meant giving up another stop on the Massif later. After 36 minutes confronting the boulder Schmitt told Parker, "I've done the best I can."

But that would prove to be more than good enough. In the months and years to come, analyses of the split boulder would testify to the violence of the impacts that formed the lunar basins. It turned out Cernan and Schmitt had both been right, in a sense; the boulder was one big breccia, but the tan-gray part had indeed been molten. It had melted because it had been subjected to forces almost beyond human comprehension. The split boulder was a kind of tableau, a geologic freeze-frame of the events that immediately followed the impact of the Serenitatis asteroid 3.8 billion years ago. Hurtling into the moon at perhaps 10 miles per second, the rocky intruder exploded

with the power of billions of H-bombs. For perhaps a hundred miles around, huge volumes of the crust were pulverized or melted and flung outward in a spray of debris. Even as ejecta flew through space, shock waves contorted the lunar crust, pushing up rings of mountains like ripples on the surface of a pond; one of the rings became the Taurus Mountains. A vast sheet of molten rock sprayed out from the blast point, blanketing the newly formed mountains, sweeping up preexisting boulders in its broiling flow. Some of the boulders were themselves breccias that had recorded earlier cataclysms. Now more debris fleeing the growing basin piled on top of the newly formed Taurus Mountains. Within minutes, it was all over. The face of the moon was scarred by a crater 450 miles across. Over time the melt sheet cooled and solidified into a layer that was visible high on the Massifs. ☾

A hundred million years later, *mare* lavas migrated upward through the fractured crust and erupted at the surface. In the Taurus-Littrow valley the last of these volcanic outpourings were accompanied by the fire fountains that created the dark mantle and the orange soil. Much, much later—"only" 800,000 years ago—a house-sized chunk from the North Massif was dislodged and tumbled down the mountainside to where two explorers now worked.

6:27 P.M.

The back room wanted a rake sample, so Schmitt headed back to the Rover; if they were going to have time for another stop on the mountain, he'd better work fast. Meanwhile, Cernan struggled up the hill so that he could shoot a documentary panorama that would include the boulder. At one point Schmitt happened to walk into the field of view, and the pictures, which would show him dwarfed by the dark mass of stone, would become some of the most famous of the Apollo program; Cernan would see them reproduced in countless books and magazine articles. Looking at them, he would also see a patch of soil on top of a low, rocky ledge that would forever bear the imprint of his hands where he had collected samples; he would wish that he had taken a moment to write his daughter Tracy's name in the dust.

From this height, for the first time, he could see all of Taurus-Littrow. Up here, he could spot all the places he and Schmitt had visited on the South Massif and the valley floor. To his left, beyond the flank of the North Massif, were the Sculptured Hills, where he and Schmitt would soon be headed. And way out on the plains, a tiny dot of aluminum and Mylar. It occurred to

Much of the astronauts' activity on the steep slopes of the North Massif took place at a collection of large boulders *(left)*. The insets at the top of the picture—frames from the Rover's TV camera—show Schmitt with his gold-plated sun visor raised *(left)*, examining a boulder *(center)*, and collecting samples with Cernan.

Cernan that he had spent his adult life working farther and farther out on a limb, from flying jets to flying in earth orbit, then leaving his home world to come within nine miles of another one, and finally returning to descend from the sky into this valley, getting outside, and, yesterday, driving miles into the distance, and over a hill, so that he could not even see his lunar module. During the sleep periods, lying in that tin can, that popped and creaked in the full-strength lunar sun, Cernan's thoughts were not about undoing the steps in that chain—getting off the moon, out of lunar orbit, and safely through the earth's atmosphere—but of what he was missing, what he could be doing instead of wasting precious hours by sleeping. As much as Schmitt, he understood how little three days was when you had a place like Taurus-Littrow to explore. And yet, they'd accomplished more than he ever expected. With the excited voice of a kid who's just climbed his first mountain Cernan said, "You know, Jack, when we finish . . . we will have covered this whole valley from corner to corner!"

In an exuberant moment at the Sculptured Hills, Schmitt does a ballet-style leap while heading toward the Rover. Moments later, the astronaut pretended to ski, as if he were at a mountain resort, not doing geology on the moon.

Schmitt deadpanned, "That was the idea."

In the back room, Bill Muehlberger said, "It's only six-thirty. We've got a long night."

A surprised Dale Jackson said, "Is that what time of day it is? I'm all screwed up. I don't even know what day of the week it is." And no wonder; everyone in the back room had been going full bore for three days. Even now, with a bonanza from the split boulder in hand, they worried that Cernan and Schmitt were cutting uncomfortably into the time reserved for the next stop. They told Jim Lovell to ask that Parker hurry them along. Lovell did so, then informed Muehlberger's team that a few extra minutes late in the moonwalk would have to be reserved for some chores at the ALSEP; he wanted to know which geology stop to take it out of. Jackson replied, "Take it out of Apollo 18."

7:49 P.M.

When historians review the scores of hours of Apollo videotapes, they will no doubt take note of Gene Cernan and Jack Schmitt's visit to the Sculptured

Encouraged by Schmitt's antics, Cernan kangaroo-hops down the slope to join his crewmate at the Rover. This scene and the one on the facing page were recorded by the Rover's TV camera.

Hills and what they did there, not with their hammer and scoop, but with themselves. To be sure, the geologic rewards were great; Schmitt discovered a small boulder which, along with some of the samples from Nansen, would reveal that the ancient crust of the moon was chemically more heterogeneous than theorists had suspected. But the exploration was just as revealing of the explorers themselves, because after three days of intense effort, they let themselves have a little fun. Having climbed high upslope to the boulder, Schmitt took an excuse to perform the geologist's time-honored ritual. He called out to Cernan, "Are you ready for this?" With one boot Schmitt shoved the rock into a lazy tumble that abruptly died. He chased after it, giving it a kick that sent dust flying but did little to move the rock. "Go! Roll!" he yelled. "Look, I would roll on this slope; why don't you?" And when he and Cernan had finished breaking off a sample, Schmitt gave a ballet-style leap and bounded downhill toward the Rover. Then, forsaking any trace of adulthood, he began hopping on two feet with arms outstretched, pretending to ski. He swung his body from side to side, sample bags flopping against his backpack. He made schussing noises. "I can't keep my edges! A little hard to get a good hip rotation!" A minute later Cernan came kangaroo-hopping down the hill after him—"*Wheee!*"—as if to affirm that the strongest emotion of being on the moon was simply elation.

10:36 P.M.

The replacement fender finally gave out during the drive back to the LM. Cernan said, "There's got to be a point where the dust just overtakes you, and everything mechanical quits moving." Lunar dust is abrasive—very abrasive, because of all the little rock fragments knocked off by micrometeorites. Cernan had only to look at the handle of the geology hammer; the rubber coating was worn through to bare metal. He and Schmitt had been religious about greasing the suit zippers and cleaning the wrist rings, to keep the suits in good working order. And by God, they were holding up, despite the abuse heaped on them. Schmitt would never lose his amazement at that; just like

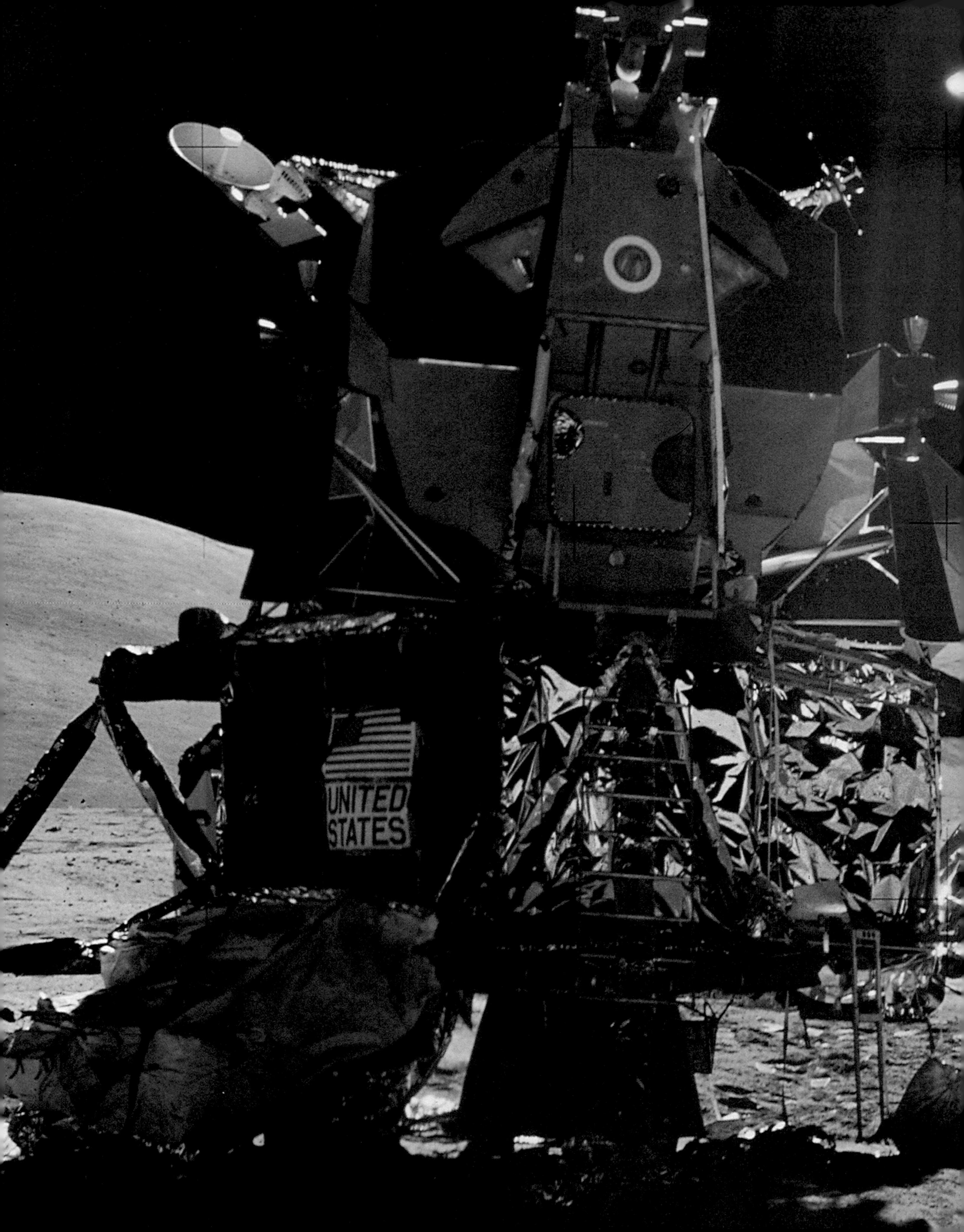
UNITED
STATES

The lunar module *Challenger* rests on the sun-washed Taurus-Littrow valley as Cernan and Schmitt prepare for their third moonwalk. Bags of unneeded gear lie under the landing leg near the center of the picture.

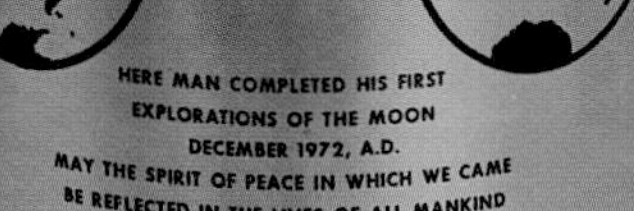

almost everything else on Apollo, those suits worked better than anybody ever expected them to. It went back to the motivation of the people who built them, tested them, and flew them. It showed what can happen when people believe that the thing they are working on is the most important thing they will do in their lives, and they don't want to be responsible for screwing it up.

No doubt those suits could have gone further, and no doubt, so could Cernan and Schmitt—but not this day. The past six hours on the surface had been the most physically demanding of all their excursions. Unlike yesterday, when they'd spent as much as an hour riding the Rover between stops, today was fast paced, with much of the work on steep slopes. At the last geology stop, on the valley floor, Cernan and Schmitt had hit a wall; they worked not for the thrill of discovery but simply to get their work done. But at the last minute, with mission control hurrying them back to the Rover, Schmitt had discovered a layer of unusually bright white soil. "Come here, Gene," he said, "we can't leave this." They made the decision, on their own, to stay longer, and mission control let them. For a time, the discovery seemed to revive them, but when the geologists tried to argue for an extension, flight director Gerry Griffin said no; he'd heard Cernan and Schmitt complain about tired hands, and wouldn't take the risk of having them end up too worn out to handle an emergency. And that was the end of it.

Griffin had been right. Back at *Challenger,* amid the litter of three days of activity, Schmitt had helped pack up the rock boxes, and it was all he could do to get his hands to work. When he discovered an unused core tube, he drove it into the ground by hand, holding onto the lander for support; then he tossed this last lunar sample in with the others. Now, all that remained was for Cernan to drive the Rover to its final resting place, where it would televise the next day's liftoff. First, however, he and Cernan would attend to a few moments of ceremony. NASA was hosting young students from seventy different countries; they were watching the moonwalk from Houston. Before the flight Cernan had the idea to distribute pieces of a rock from Taurus-Littrow to museums around the world, and NASA planned to tie it in with the visit of these children. And so he and Cernan stood before the TV camera while Cernan, holding a rock, explained.

"It's a rock composed of many fragments, of many sizes, and many shapes, probably from all parts of the moon, perhaps billions of years old . . . that have grown together to become a cohesive rock, outlasting the nature of space, sort of living together in a very coherent, very peaceful manner. . . ." Cernan would look back wishing he'd had more time to prepare his remarks. ". . . We hope this will be a symbol of what our feelings are . . . and a symbol of mankind: that we can live in peace and harmony in the future.

"And now—let me bring this camera around—" Cernan grabbed the TV camera and pointed it at *Challenger*'s front landing gear, where a plaque commemorated the final lunar landing. Schmitt's contribution to that plaque had been to add, below the customary pictures of earth, a small picture of the moon, showing the places where men had landed. But Schmitt had not been successful in another hard-fought effort, to change its wording. He had wanted something to reflect what he still firmly believed, that even if no one came back to the moon for twenty years—let those explorers include Joe Allen's little boy and other children of his generation—the return was

Upon returning to *Challenger* from the final traverse, Cernan unveils a plaque *(inset)* fastened to the LM's front landing leg. The inscription on the plaque reads: "Here man completed his first explorations of the moon, December 1972 A.D. May the spirit of peace in which we came be reflected in the lives of all mankind."

inevitable. Instead, the message spoke of finality. Now Cernan removed the cover and read the message.

"Here man completed his first explorations of the moon, December 1972 A.D. May the spirit of peace in which we came be reflected in the lives of all mankind." Schmitt listened to his commander's words. Within his suit he heard the reassuring whir of cooling pumps in his backpack, a sound he had almost come to take for granted. Now Cernan was saying, "This is our commemoration that will be here until someone like us, until some of you who are out there, who are the promise of the future, come back to read it again and to further the exploration and meaning of Apollo." Schmitt's fingers were raw within his gloves. His arms ached with the exertions of a full working day on the surface. And as he listened, he realized: *It's over.*

11:34 P.M.

After loading the rock boxes and other cargo into the LM—he had to climb up the ladder several times to pass the massive stowage bags to Schmitt—Gene Cernan stood alone on the moon, his space suit covered with grime. He was somewhat out of breath. Now he spoke, trying to cut through the fatigue, to imbue his words with energy, to speak for the history books. "Bob, this is Gene. As I take man's last steps from the surface, back home for some time to come—but we believe not too long into the future . . ." His words were interrupted by the sound of his own breathing. ". . . I believe history will record that America's challenge of today has forged man's destiny of tomorrow." As he spoke, Cernan took a last look at the bright and barren wilderness all around him. "And as we leave the moon at Taurus-Littrow, we leave as we came, and, God willing, as we shall return, with peace and hope for all mankind." He glanced at the earth, high in the southwestern sky, an unmoving, silent witness to a voyager about to begin a long journey back. Once more, he spoke: "Godspeed the crew of Apollo 17."

FRIDAY, DECEMBER 15
LATE EVENING,
HOUSTON TIME
ABOARD *AMERICA*,
IN LUNAR ORBIT

Gene Cernan would never have believed that he could feel so comfortable with his neck so far out. It had been more than a day since *Challenger* had blasted off from Taurus-Littrow, carrying him and Schmitt into orbit and then, like riding on rails, back to a very happy Ron Evans. In the last phase of

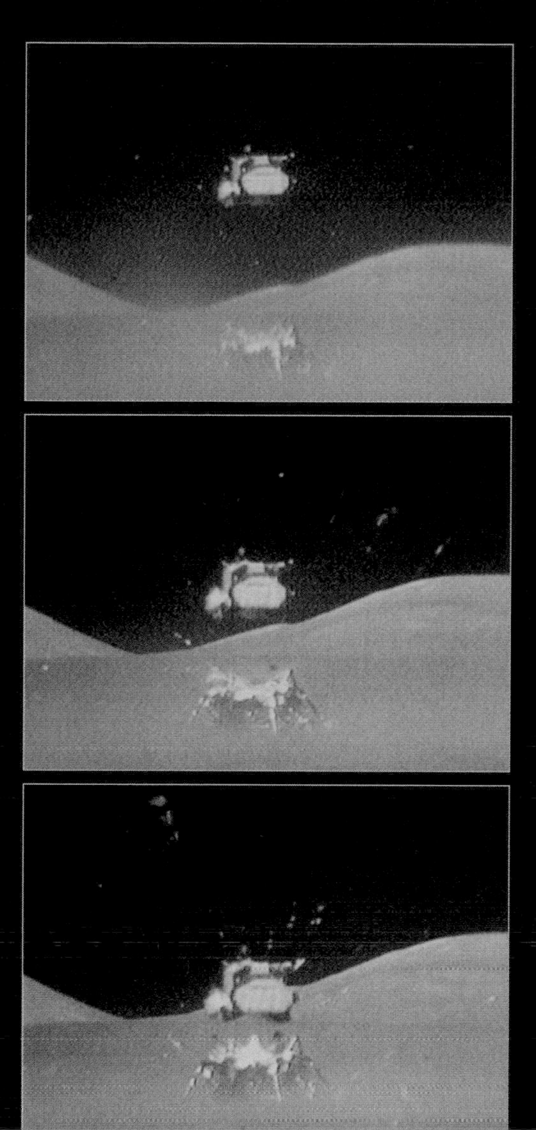

the rendezvous Cernan flew his craft in tight formation with the command module, as *Challenger* danced like a fighter plane under his control. It was his moment of exultation at a mission accomplished; it was his victory roll. With 60 percent of his maneuvering fuel still untouched Cernan would have loved to fly beautiful orbital-mechanics circles around Evans, but he did not. There could be no distractions. Minutes later he held the ascent stage still while Evans approached, ever closer, until the cabin filled with the welcome ripple-bang of twelve docking latches snapping shut. And Cernan felt that he had come home. Inside *America,* he couldn't believe how good it felt just to get out of his filthy pressure suit, strip off his long johns, and to wet a washcloth and wash the sweat and the dust off his body. Nothing would get his fingernails clean—the dust had penetrated deep into the quick; it would be weeks before it grew out.

No matter; Cernan felt like a new man, even though he and his crew were spending two more days in lunar orbit. During training, Cernan hadn't been in favor of these extra days in orbit, but Evans and Schmitt had been so enthusias-

With a spray of foil insulation *(below)*, *Challenger*'s ascent engine lifts Cernan and Schmitt from the lunar surface *(above)*. Televised from the Rover on December 14, the departure felt to the astronauts something like a ride on a very quiet, very fast elevator.

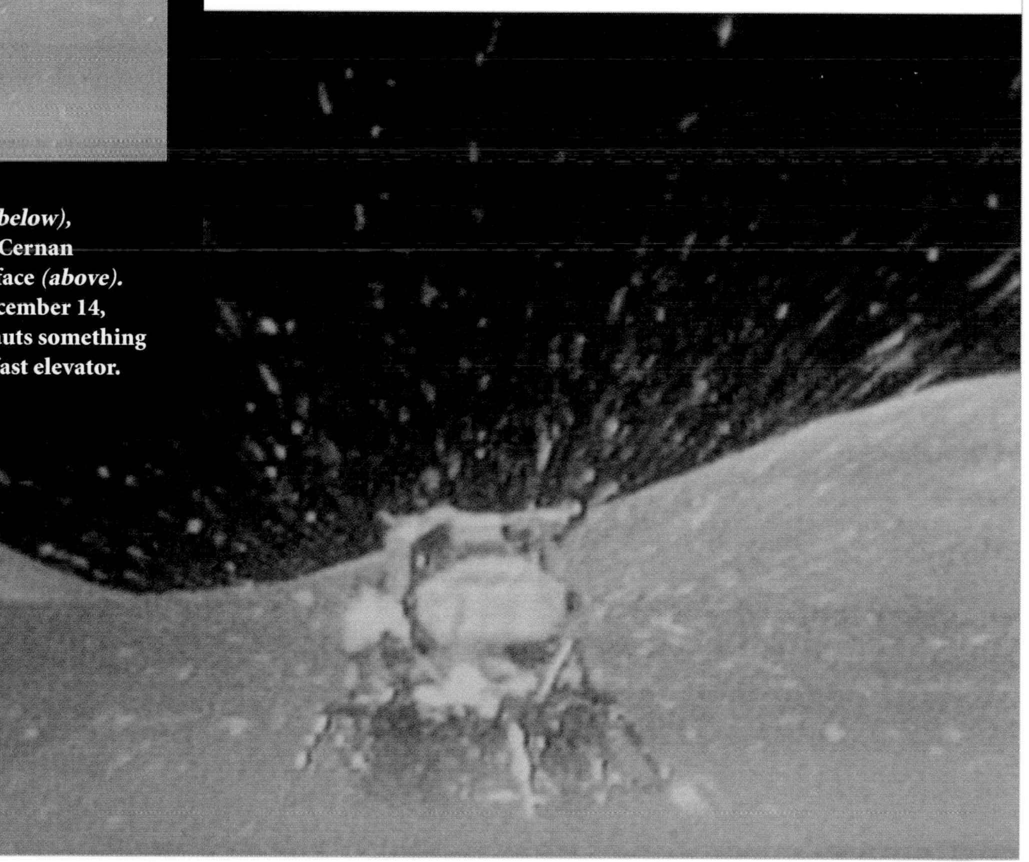

Gene Cernan, in his clear bubble helmet, peers intently through one of *Challenger's* triangular windows as he guides the ascent stage toward a docking with the command module *America*—and a reunion with its pilot Ron Evans.

tic about making more observations that he'd backed off. And now that he was here, he marveled at his own ease. He could settle in and soak up the view. When it came time to leave, then he would worry about the leaving.

Cernan was proud of Apollo 17. With 75 hours on the moon, 22 of them outside, and nearly 19 miles total distance covered, this mission had made its mark on the record books. Not a single major problem with the spacecraft; arguably the smoothest and most successful flight of the series. And he had accomplished the goal that he had set for himself almost a decade ago. He could say to himself what only five other men could say: He had steered a spaceship to a landing on the moon.

That evening, when *America* was on its 66th circuit of the moon, Jack Schmitt saw his first earthrise. Looking at that bright crescent, Schmitt sensed the same fragility his predecessors had seen. It had nothing to do with the earth's appearance, he would say later; it stemmed from its monumental isolation in the blackness. The earthrise brightened Schmitt's dark mood for a moment. Down on the surface, just before liftoff, he'd felt the first stirrings of sadness at the end of Apollo. Though he felt proud of his accomplishments on the flight, he had never seen his getting to the moon as a personal triumph; he'd always felt that he and Cernan were just lucky enough to implement the triumph of others. Apollo 17 represented the last acts of lunar exploration that would be done by members of their generation.

But the worst moment came later, just after the docking. The vibrations hadn't even died down when Gordon Fullerton called up and said, "I'd like to take a minute of your time here to read the following statement by the president of the United States of America." Schmitt was annoyed; they had work to do. But Fullerton went ahead and read the message.

"As the *Challenger* leaves the surface of the moon, we are conscious not of what we leave behind, but of what lies before us." Even in Fullerton's quiet, even tone, there was Nixon's cadence. "The dreams that draw humanity forward seem always to be redeemed, if we believe in them strongly enough and pursue them with diligence and courage." Ironic words, thought Schmitt, considering the fact that Apollo had been conceived under Kennedy and was ending under Nixon. Then Fullerton read the sentence that sank Schmitt's spirits. "This may be the last time in this century that men will walk on the moon, but space exploration will continue . . ." Schmitt barely heard what followed. He couldn't believe his ears—*the last time in this century?* He hated the words—hated them for their lack of vision. These words, from the leader of the nation! Even if Nixon really believed them, he didn't have to say so in a

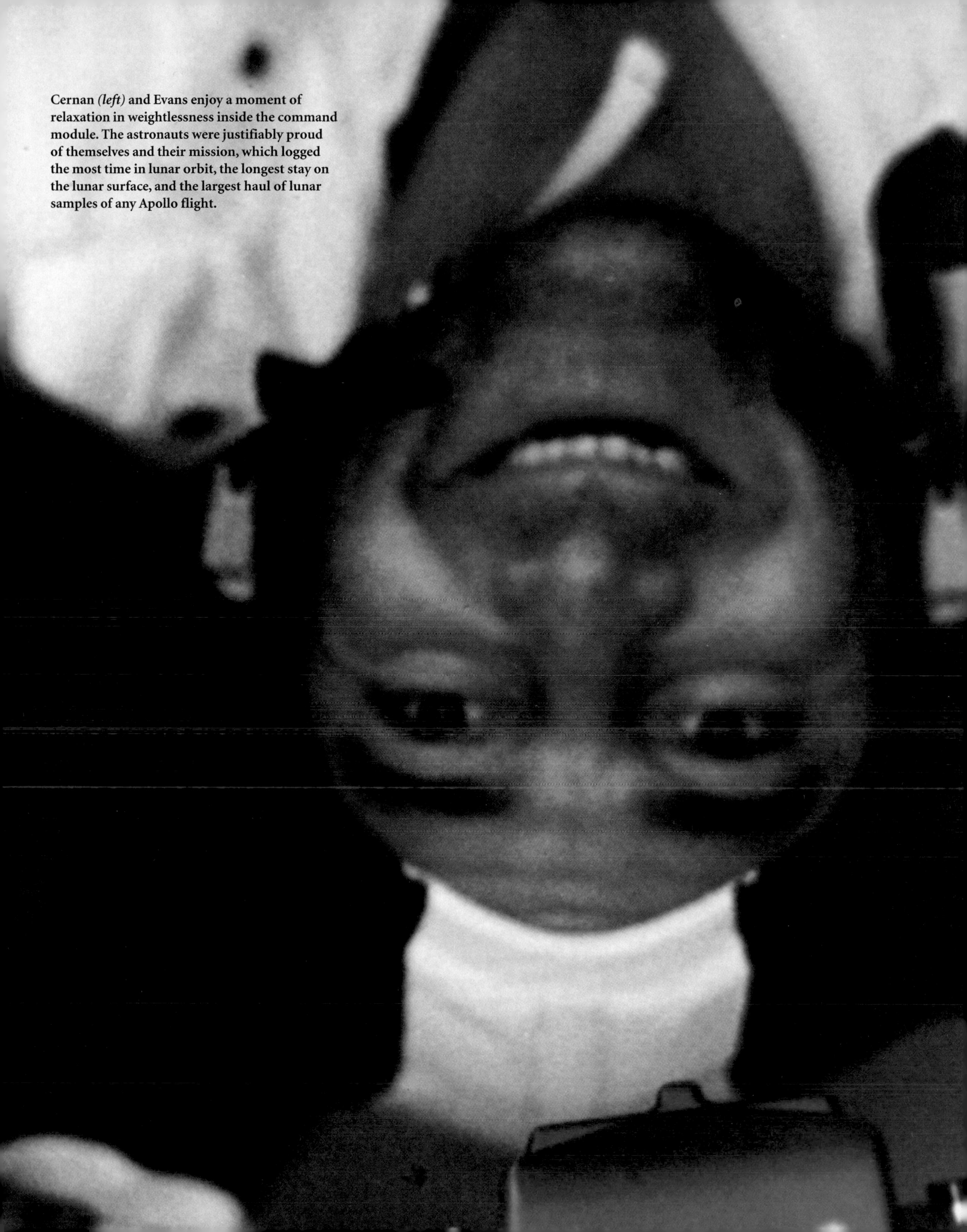

Cernan *(left)* and Evans enjoy a moment of relaxation in weightlessness inside the command module. The astronauts were justifiably proud of themselves and their mission, which logged the most time in lunar orbit, the longest stay on the lunar surface, and the largest haul of lunar samples of any Apollo flight.

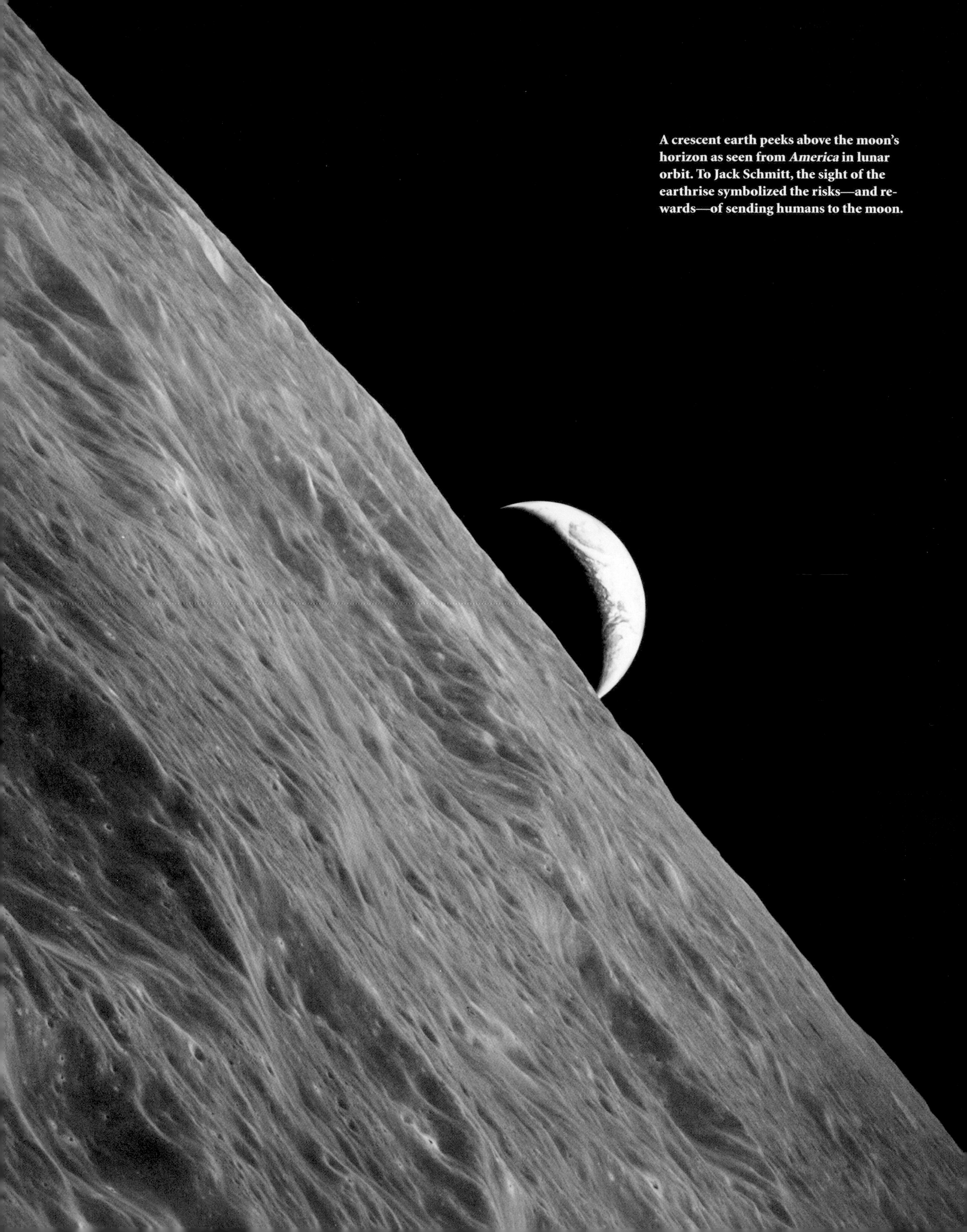

A crescent earth peeks above the moon's horizon as seen from *America* in lunar orbit. To Jack Schmitt, the sight of the earthrise symbolized the risks—and rewards—of sending humans to the moon.

public statement, taking away the hopes of a generation of young people. Schmitt was furious that in the moment of triumph, he had been jolted out of the work of the mission to listen to a statement like that. He would fume silently about it for the rest of the flight.

SATURDAY, DECEMBER 16
5:54 P.M., HOUSTON TIME
9 DAYS, 21 HOURS,
1 MINUTE MISSION
ELAPSED TIME

When the SPS engine shut down, and a perfect Transearth Injection burn was behind them, Cernan's crew broke out the television camera. When *America* regained contact with earth, they broadcast a view of a huge and nearly full moon. Part of the far side was still visible, including Tsiolkovsky. One of the reasons Schmitt had pushed for a landing there, back in 1970, was that he had hoped it might reignite enough public interest to prompt NASA to reinstate Apollo 18 and 19. As he panned the TV camera across the face of the moon, he captured not only places he had hoped astronauts would go—the lunar poles, and Tycho—but some of the places where they had been: Tranquillity Base, and Hadley, the Descartes highlands, and a valley at the edge of the Sea of Serenity that would forever be known as Taurus-Littrow.

It was television that set Apollo apart as an exploratory venture. Schmitt could still remember the December night four years ago when he and engineer Jack Sevier had stayed up trying to figure out how Frank Borman's crew might televise a picture of the earth. They ended up telling them to use every filter in the photography kit together, taping them in front of the lens. The fix was a little inelegant, but it worked; that day, humanity saw its home on live TV for the first time. Eight months later, on a hot July night, the pictures were transmitted from a patch of dusty plains on the Sea of Tranquillity. Something extraordinary happened when those fuzzy black-and-white images appeared on television screens around the world: For the first time in the history of exploration, the human species—the developed world and the developing nations—participated in the moment along with the explorers themselves.

Now Schmitt was returning to that earth, having seen it from a great distance. The data held in his mind would be shared with his earthbound colleagues in a trailer at the space center in the days before the Christmas holiday. Then, for a time, he would enter into the debriefing of the self, and his thoughts would return to the sight of that small and lovely crescent beyond the alien shore. The ability to witness an earthrise, he would note, could happen only if humankind took the well-planned but still significant risk of

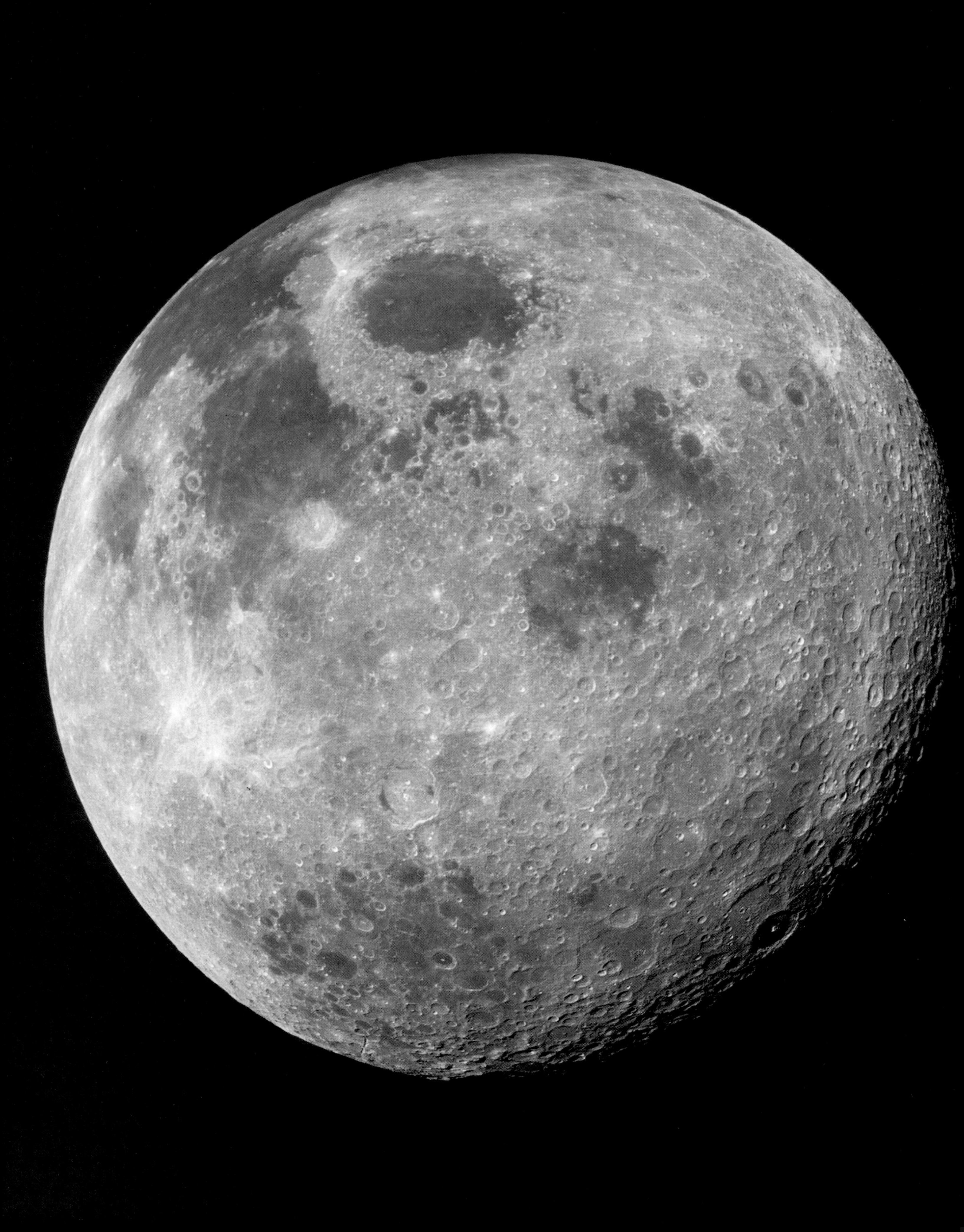

A nearly full moon hangs in the blackness of space as the astronauts begin their return to earth. Near the upper left edge of the moon are the Seas of Serenity, Tranquillity, and Crisis. Along the lower right edge are the highlands of the lunar far side.

sending three of its members into an environment utterly foreign to the one in which they had evolved—that is, lunar orbit—with no guarantee of getting those people back. That commitment, Schmitt would come to believe, marked a turning point in human evolution. Human beings had become a space-faring species. What they would do with that new status was still locked in the unknowable future, but Schmitt firmly believed that space exploration would dominate the future of humanity. ☾

And what of that humanity, living into another day, unmindful of the leap it had made from the turning earth? In particular, what events were drawing attention in a country called the United States? In the headlines, Henry Kissinger had reported that serious problems had brought peace talks between the United States and North Vietnam to a halt. The United Mine Workers had elected a new president who pledged to clean up the union. A blizzard had dumped 28 inches of snow on northeastern Ohio, blocking highways and closing airports, with another foot expected before the storm moved east. In a Kansas City hospital, the man who had overseen the final step in the transformation of a nation of revolutionaries, then pioneers, and then immigrants, into a superpower, was battling lung congestion, heart problems, and kidney malfunction. Harry Truman was dying. And in a space ship called *America*, three men were coming home from the moon.

SUNDAY, DECEMBER 17
2:37 P.M., HOUSTON TIME
MISSION OPERATIONS CONTROL ROOM, MANNED SPACECRAFT CENTER

Stu Roosa sat next to the Capcom, looking at the big screen at the front of mission control. The image was stark, and for the uninitiated, like some of the visitors up in the viewing gallery, it must have looked very strange. It was mostly dark, except for a big white object that looked something like a snowman, or perhaps a mummy, lying on its stomach. But Roosa knew exactly what he was seeing. The figure was Ron Evans. He was floating in the void 180,000 miles from earth, beginning his space walk to retrieve the photographic film from the side of the service module. Roosa listened to the transmissions from Apollo 17. He heard a whooshing sound—that was oxygen flowing into Evans's suit from the umbilical. And he heard Evans's exuberant voice: "Hot diggety dog!" There was nothing complicated about Ron Evans. He'd worked just as hard backing up Roosa for Apollo 14 as he had for his own flight. Now he was in his element. "Talk about being a spaceman, this is it!" On the screen, Roosa could see him waving. "Hello, Mom! Hello, Jan! Hi, Jon! How you doing? Hi, Jaime!"

As Ron Evans takes a space walk to retrieve cassettes of scientific film from the service module's mapping and panorama cameras, his gold-plated sun visor reflects his life-support umbilical and the command module *America*. Behind him appears the crescent earth, approximately 154,000 miles—two days—away.

There wasn't anyone in that control room who truly understood how Roosa felt, watching this. He had trained to make that space walk himself, dozens of times. With the lack of flights, Deke Slayton started putting experienced astronauts on the backup crews to save the extra time, money, and effort that would have been required to train rookies. When Roosa and Ed Mitchell came off Apollo 14, they were assigned to back up 16, with Fred Haise as the commander. Before Apollo 14, Mitchell had seen the assignment coming and had balked, but Slayton made it clear that he'd better go along or he wouldn't fly. But there was no resistance from Roosa. He'd wanted the assignment. There was always the chance that something would happen, and he'd get to go—though he would have had some very mixed feelings about flying if it meant Ken Mattingly were grounded again. Still, on the day Apollo 16 was launched he was standing with Fred Haise, and Haise was talking about how hard it was to train for more than a year, and get all cranked up to fly, and then just stand there and watch somebody else go instead. If Roosa said anything in response, it wasn't necessary. Anyone who'd ever done time on a backup crew knew what he was talking about. It was right around that time that the stamp affair broke, and in the shakeup that followed, Scott,

Worden, and Irwin were taken off the Apollo 17 backup crew, and in their place Slayton assigned Roosa along with John Young and Charlie Duke, just back from their world tour. And Stu Roosa, veteran astronaut, found himself serving on two backup crews in a row. The three of them joked about being the only all-Southern crew. They grew mustaches, just for the hell of it, and they trained for a mission they knew they would probably never fly. By the time launch day rolled around Roosa figured he had more time in the command module simulator than any other astronaut. He was ready to go back to the moon—more than ready. His best friend Charlie Duke felt the same way, but he knew nothing was going to keep Jack Schmitt from getting to the moon, short of a broken leg—and even then, Duke figured, NASA would probably postpone the mission. Roosa had been more hopeful. He wanted a chance to do it again, and do it even better—if that was humanly possible. He'd take more photographs; he'd make more observations. And this time he wouldn't let so much of the experience get away from him. He'd be more relaxed; he'd have time to collect his thoughts. He'd *enjoy* it a little more. And he wanted to make that space walk, 180,000 miles from earth. But that experience, like a second trip to the moon, was now forever beyond his reach.

"Okay, put the old sun visor down now. I see what Charlie Duke meant. Man, it's dark out here . . . Whoops, come back here, little cassette . . . Okay, I'm coming back in."

●◐◐○○○◑◑

A few days later, when Apollo 17 splashed down, the control room was so packed with people that it was hard to walk around. It seemed every flight director and flight controller who'd ever worked on Apollo had been drawn to this room to see the last mission end. And Roosa was there, seated next to the Capcom, along with Deke Slayton. He watched quietly as the command module descended under three perfect parachutes, while around him the room broke into applause. By the time Cernan, Evans, and Schmitt were being hoisted up to the helicopter, cigars were lit. Everyone was clapping and shaking hands. Minutes later the big screen showed Cernan and his crew on the carrier deck, making speeches.

There were probably as many emotions in that room as there were people. Chris Kraft would say later that he was absolutely relieved when he saw the parachutes; they were getting out of this risky business at just the right time, and with a great finish. One of the flight controllers would liken his own wistful sense of accomplishment to the feeling the architects of the pyramids must have had when the last one was finished. By nightfall there were

A spray of seawater marks *America's* splashdown in the Pacific on December 19, as a delighted Jan Evans watches with daughter Jaime (*above, left*), son Jon, and her sister, Marian Bell.

splashdown parties in full swing all around the space-center communities, but there wasn't the same all-out feeling of celebration they'd had in 1969.

In 1973 Roosa was assigned to the Space Shuttle program, and for the next few years he went to meetings and saw the program fall further and further behind. By 1976 it was clear that the first flight would probably slip well beyond its 1978 target date. And Roosa knew he wasn't first in line; by the time he had a chance to fly again, it might be more than a decade after Apollo 14. He'd be a spacecraft commander, but he'd also be forty-seven years old. But if he was going to leave what Ron Evans called "the best job in the world," what would he do? He thought about going back to the air force, but he was told there was a lot of anti-astronaut sentiment there, because the stamp affair had been done by an all-air-force crew. So Roosa decided to leave NASA and try his hand at business. He'd been contacted by a corporation that wanted to expand its operations in the Middle East. They had offered him a vice president's position in Athens. It sounded challenging, and the kids would enjoy the change. But Roosa never really felt that he'd found something to replace what he did from 1966 to 1972. Someone had called it Mankind's Greatest Adventure; what could Roosa add to that? On the day he decided to leave, he called George Low, and they talked about his plans; Roosa said he was really going to miss NASA, but that it was time to move on. Then Low was quiet for a moment, and he said, "You know, there will never be another Apollo in anybody's life."

A half-earth gleams above Cernan and the U.S. flag that he and Schmitt planted shortly after beginning their first moonwalk. Because the Taurus-Littrow valley is located relatively far from the geographic center of the moon's near side, the earth was closer to the horizon there than at any other Apollo landing site.

NASA

Cernan kneels while attempting to extract a deep core sample during the first moonwalk. Rising beyond him, the Wessex Cleft marks the junction between the North Massif on the left and the Sculptured Hills on the right. Though they appear to be nearby, the peaks are actually more than two miles away.

At Camelot crater, Schmitt holds the lunar rake, which is full of cobble-size rock samples. The object on the ground between him and the Rover is an explosive charge to be detonated from earth after the astronauts had left the moon. The resulting shock waves would be used to analyze the subsurface structure of the landing site.

Precious and distant, the earth hangs above a boulder near Nansen crater. Cernan, who was taking photographs to document the area, stole a moment to snap this portrait of his home planet. As commander of the final lunar landing mission, Cernan became the last man to walk on the moon.

THE AUDIENCES OF THE MOON

In 1975, during a trip to Nepal with his wife, Joan, Stu Roosa visited a school to give a talk about his flight to the moon. Afterward, there were mysterious questions: "Who did you see?"

Roosa answered, "There is no one there." A murmur went through the place. Again the students asked what he had seen. Roosa was adamant: "There is nothing there. Not even wind. There is nothing."

Later, after the Roosas had gone, a teacher told the children, "You mustn't listen to him. He's wrong."

The Roosas were distressed when they learned that some Nepalese believe the spirits of their ancestors reside on the moon. Roosa had essentially told them there was no heaven. Joan wished the American government had briefed them better for the trip.

One day at the hotel, Joan went to get her hair done. The woman giving her a shampoo said, "You're married to the astronaut, aren't you?"

"Yes, I am."

"You know, you're married to a god." Joan laughed it off. But when she and Stuart went to that evening's gathering, all along the mountain road children were kneeling with candles, in an act of reverence: a god had come to visit.

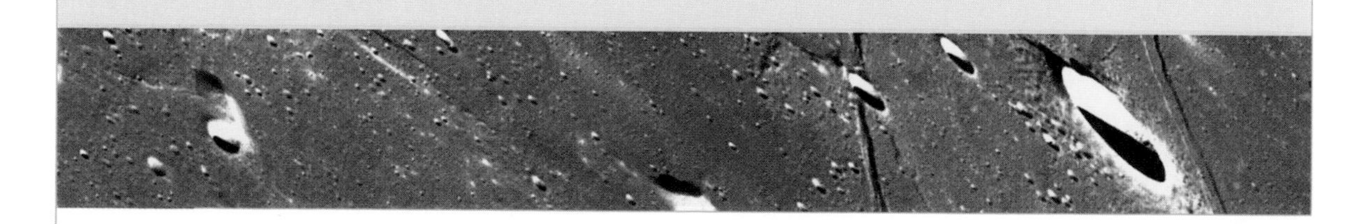

Pete Conrad, 1994

In the mid-1970s, a series of television advertisements for the American Express card featured notables whose faces were unknown to the public. Two of them were William Miller, who ran for vice president in 1964, and Mel Blanc, who did the voice of Bugs Bunny. And in one of the great ironies of our age, the producers chose a man who had been to the moon. On TVs across America, there was Pete Conrad, grinning his gap-toothed grin, saying, "Do you know me? I walked on the moon."

For Conrad, the ad changed things—for a little while. "Up until the time I did that," said Conrad, "I could go *anywhere,* and nobody knew me. Alright? I did that American Express commercial, and while that sumbitch was running, I couldn't take a leak—I remember trying to take a leak in Philadelphia, and a guy standing there says"—Conrad lurches into a crazed persona—"*I know you!*" Then the ad stopped running, and Conrad's sudden fame evaporated as quickly as it had come.

Today the moon voyagers are scattered across the country, and most of them are as anonymous as Pete Conrad. Many, like him, became corporate executives. And most agree with the credo Conrad stated one day, sitting in his office at McDonnell Douglas in Long Beach, California: "Don't ever look back. I'm serious. What's the point of looking back? I've already been there. How about some old guy sitting around telling you about how he played football for good ol' Yale University and was the world's greatest quarterback—that's all the bastard ever talks about. Okay?" To Conrad, the scene was so awful that no further explanation was necessary.

From time to time, however—usually for the benefit of an interviewer—Conrad did look back, and even among astronauts he was known as a great spinner of space yarns, or Lies and Sea Stories, as he called them. But Conrad kept the promise he made to himself on the day he was selected as an astronaut in 1962. Going to the moon, he said, didn't change him.

But that isn't what people want to hear, said Conrad. "They've got some preconceived notion that I should tell 'em I was frightened, or I was awe-inspired, or I saw the Lord—or . . . I don't know." The truth sounds unbelievable: when he was on the moon his strongest feeling was that it was the right place to be at the time. "That just shuts the door." So does the thought of a man who has been to the moon and who, today, does not look at it. "They *know* I'm *lying.* They say, 'Don't you go out and look at the moon?' And I honestly don't."

When Pete Conrad looked back on his spaceflight career, the high point wasn't his lunar landing: it was the rescue of the Skylab space station in 1973. The station's outer shield, which protected it against heat and micrometeorites, was torn off during launch, taking one of its power-producing solar panels with it. Getting power to Skylab depended on freeing the remaining wing, which was lashed down by debris. Conrad's crew, who had been preparing for a month-long mission aboard Skylab, was now faced with carrying out a demanding repair. Arriving at the stricken station, they sweltered in desertlike heat for days until they could erect a makeshift sunshield. Two weeks into the flight, Conrad and his crewmate Joe Kerwin made a space walk and, with some difficulty, freed the stuck solar wing. Long afterward, Conrad looked back with a healthy appreciation of the risks he took.

"My life was a lot further out on the line . . . on Skylab than it was on the moon," Conrad said. "That taxed me *personally,* put everything that I had spent my whole life . . . learning how to do, on the line. . . . Going to the moon was basically a nice, routine flight after the lightning. We didn't have any trouble after that. On Skylab we didn't know whether we [would leave] or stay for fourteen days." In 1978, Congress recognized the success by awarding Conrad the newly created Space Medal of Honor.

"Everybody thinks I got the Space Medal of Honor because I went to the moon. I say, 'No, it was for Skylab.' They say, 'Oh, Skylab. Yes. What was Skylab?' " It's the moon that people want to hear about, and like all his colleagues, whenever Conrad was introduced as one of the twenty-four men who went there, the question he was almost always asked was, What was it like? And he gave the neat, two-second answer he developed long ago: "*Super!* Really enjoyed it."

"Space changes nobody," Stu Roosa said flatly in 1990. To be sure, Roosa talked about the impact of seeing the earth from deep space, and the lunar

Ed Mitchell, 1994

landscape. You can't see those things and not be affected. But it didn't change him. "You bring back from space what you bring into space," he said.

Roosa's words are ironic, considering the fact that one of the two men who rode with him on Apollo 14 says his own perspective, his life, even his very sense of being, were profoundly altered by the voyage. He is Ed Mitchell. Since 1971, Mitchell has tried to understand what happened to him inside the command module *Kitty Hawk* heading toward earth, when he saw the universe revealed in a flash of understanding. Now a freelance management consultant in Boca Raton, Florida, Mitchell talks with quiet intensity about his consciousness-changing experience in space, and the journey that followed. ☾

When Mitchell came back from Apollo 14, the other astronauts heard about his attempt to transmit his thoughts through space. Some, he says, even came by his office to talk about it—when no one was watching. But none of them knew that Mitchell had returned to earth a changed man; even he did not really comprehend his experience.

"I didn't know what feelings were," Mitchell says of himself as a returned moon voyager. "People used to ask, 'What did it feel like to be on the moon?' I didn't know what it felt like! I could tell them what I did, and what I thought, but not how I felt. It pissed me off." But Mitchell realized that answering the question was a key to unraveling his revelation in space. He sought out two researchers in consciousness, Jean Houston and her husband, Bob Masters, and through them, relived the experience via hypnosis. From there, Mitchell began to gain understanding. He spent the next fifteen years piecing together a new truth.

In 1973, he founded the Institute of Noetic Sciences in Palo Alto, California, devoted to the scientific study of consciousness. Even after leaving daily operations at the institute in 1982, Mitchell continued his quest for scientific explanations of consciousness. He helped arrange experiments to test the claims of psychic Uri Geller, whose spoon-bending and other feats of telekinesis made him a talk-show celebrity. He has also delved into the work of such avant-garde scientists as physicist David Bohm, who proposed a kind of consciousness at the level of subatomic particles. Like Bohm, Mitchell believes that consciousness is an attribute of everything in existence, animate and inanimate.

Today, Mitchell believes that telepathy, clairvoyance, intuition, and other psychic phenomena have a single explanation: "It's just *information.*" Information, Mitchell says, is the other half of the universe, the intangible complement to matter. An information field pervades space, like a cosmic data bank, that records, among other things, the accumulated experiences of all matter, including living organisms throughout history. Mitchell believes that a person who is sensitive to this information—which exists, he says, as a form of energy—can experience anything from clairvoyance to telekinesis. When he had a shift in consciousness on the way home from the moon, he says, he was tuning in to the data bank. Why didn't the other astronauts experience the same thing? They *did,* according to Mitchell, whether or not they express it. "The informational *input* is the same. But what is [actually received] is shaped by our level of awareness, and by the filter we call belief system." Mitchell believes he has found the resolution between science and spirituality that he hungered for as a teenager: "To me, divinity is the intelligence existing in the

Jim Irwin, 1989

universe." That universe, he adds, is a learning, growing, changing organism, like the human beings who strive to understand it, and who have only begun to explore it.

"It is my joy to be with you tonight, to share the love of our Lord and Savior, Jesus Christ," Jim Irwin said in 1988, standing before a packed meeting hall in Colorado Springs. After a series of major heart attacks, he appeared thin and almost frail compared to his astronaut days. But his eyes were bright and penetrating, revealing an unmistakable inner strength and determination.

Information, Mitchell says, is the other half of the universe, the intangible complement to matter.

"I come to you as a former astronaut," Irwin continued. A wry smile crossed his face. "I feel like the ancient astronaut. The shrinking astronaut. But I'm glad that I'm still alive and that I can be here." This is how Jim Irwin spent the last two decades of his life, sharing his faith, testifying to the love of God that he discovered, he said, on the moon. Irwin embraced the task of recounting his experiences again and again, reciting stories he had long ago committed to memory. He talked about the white rock he spotted at Spur crater, presented on its own pedestal as if on an outstretched hand. Holding up a plastic model of the rock—there is almost nothing left of the real one, he would say, which had been cut up and sent to scientists around the world—he talked of the real impact of his trip to the moon, his realization that God is alive and available everywhere in the universe. ☾

Back on earth, that awakening caused Irwin's life to change dramatically. Urged by his wife to heed this new spiritual call, supported by family friend and minister Bill Rittenhouse, Irwin left the astronaut corps and founded the High Flight ministry in 1972. The timing was significant; the stamp affair had just become news. By publicly admitting that he had made a mistake, Irwin took himself off the pedestal reserved for conquerors of the heavens; at the same time he began a life that could not be more different from the one he led as a test pilot. The quiet, introverted engineer now faced a life of public speaking, something he said he dreaded but grew to enjoy. One month he was in India, visiting polio-stricken children; another, he was witnessing in Eastern Europe. He gave his testimony in churches across the United States. Someone nicknamed him the moon missionary. ☾

It was hard for some of Irwin's astronaut colleagues to understand his transformation; they would joke, "I don't know what happened to ol' Jim up there. . . ." Irwin had expected that response, but for his own part, he had no doubts about why his life took the path it did. "God had a plan for me, to leave the earth and to share the adventure with others, so that they can be lifted up." Going to the moon, he said, "prepared me for a role of greater service." To attend one of Irwin's appearances was to get a sense of how many lives he touched. In 1989, after sharing his faith on the campus of a small Baptist college near Worcester, Massachusetts, Irwin was mobbed by admirers and signed autographs for nearly an hour. That night the organizer of a dinner given in Irwin's honor by the local Hispanic community spoke of her dream to bring Irwin to Puerto Rico, so that the children there could hear how he reached for the stars, and know that they could do the same.

Irwin managed to find an element of exploration in his new life. Beginning in 1973 he participated in several expeditions to Turkey's Mount Ararat to search for the remains of Noah's Ark. In 1982, Irwin took a bad fall on Ararat's rocky slopes, almost losing his life. In the summer of 1986, near his Colorado Springs home, Irwin suffered a near-fatal cardiac arrest. Only weeks later he was recovering at home, pedaling an exercise bike and talking about going back to Turkey to continue his quest.

"I realize that my time is running out," he said calmly, looking off toward the mountains. "I want to take care of myself as best I can so I can stick around and be of help. I just hope that when the Lord's finished with me that I go quickly."

In the years that followed Irwin continued a grueling schedule of appearances, cutting down to two a day after his doctor warned him to slow down. In August 1991, Irwin died following a heart attack while in the Colorado mountains he loved.

If you talk to the moon voyagers, you will find that most of them think about the experience the way Bill Anders does. He doesn't describe the kind of life-changing impact that came to Jim Irwin and Ed Mitchell. He does say that the experience of being one of the first three men to circle the moon—and in particular, seeing the earthrise—broadened his perspective forever. It also made him something of an oracle, for a time. On the banquet circuit after Apollo 8, people would ask him and his crewmates about the chances

Bill Anders, 1994

for world peace and how the stock market was going to behave. "Pretty soon," Anders laughs, "it's easier to answer than to explain we don't know anything about it."

That wasn't the only thing Anders didn't know about when he left NASA. Like his colleagues, he was highly trained for the business of spaceflight—and untrained for real dollars-and-cents business. But he was clear about one thing: he didn't want to build a new career on his status as a moon voyager. It did open doors, however; in 1969, the Nixon White House appointed him as executive secretary of its National Aeronautics and Space Council. He went on to head the Nuclear Regulatory Commission, making use of his pre-NASA experience as a nuclear engineer. His performance in that job won him a stint as ambassador to Norway. In 1977, after eight years of government jobs in which he learned, among other things, "how to keep out of the way of Machiavelli's knife," Anders was recruited for an executive vice president's post at General Electric, which he called his "business boot camp" experience. In 1984 he was hired by Textron to run their aerospace and commercial divisions. Within a few years Anders was Textron's chief operating officer, earning more than a million dollars a year.

One day in 1987 Anders reflected on the challenges he and his colleagues faced in making second careers. "You're given unrealistic opportunities that you wouldn't have had otherwise, and measured by an unrealistic yardstick." The title of ex-astronaut may open some doors, Anders says, but once inside, "you really have to perform."

For Mike Collins, that meant a year as the assistant secretary of state for public affairs, a job that taught him, in his words, "how to operate in the strange, semihysterical environment" of official Washington. Thus prepared, he became the first director of the Smithsonian's new National Air and Space Museum, overseeing its design and construction. It opened ahead of schedule, a few days before the Bicentennial in 1976, and within budget—an almost unheard-of set of circumstances. Collins's hope that it would be the most exciting museum in the world was borne out by attendance figures: in three weeks it drew a million visitors; by the end of the year that figure had risen to 10 million.

For Jack Schmitt, who refuses to see his moon flight as the high point of his career—"The whole thing was a continuum for me," he says, "a steadily rising level of experience and no dropoff at the end"—recognition as a former astronaut helped him run for a Senate seat in 1976 against senior Democrat Joseph Montoya. "All the astronaut thing did was balance that ledger. He had seniority, I had astronaut. I could get as many speaking engagements as he could. But it didn't win the election. What won the election was, the Democrats were mad at Montoya."

Schmitt hoped he might bring "a different intellectual base to the discussion" of issues in the Senate, but his term was a frustrating and disillusioning experience. He felt like more of an outsider in the Senate than he had in the astronaut corps, he explains, because "I was a Republican, scientist, technologist. Interested in issues of the future, in addition to those of the present. And there wasn't anybody else there interested. There still isn't. . . . The Congress of the United States is not fertile ground for working those kinds of problems." Schmitt lost a reelection bid in 1982.

Meanwhile, for Bill Anders, success came at a price. He was on the road constantly; his only moments of relaxation were found aboard his sailboat, *Apogee*. "Money isn't everything," Anders said in 1989. "*Sailing* is everything." That year, when Anders was seriously thinking of retiring, he was recruited to head General Dynamics, one of the largest defense contractors in the country. By 1992, however, the Cold War was over, and Anders was in charge of guiding General Dynamics through the turbulent transition. After a controversial recovery effort that involved massive layoffs and a lucrative incentive program for the company managers, General Dynamics made a dramatic turnaround. GD's stock price more than quadrupled, and the company returned more than $2 billion to lenders and shareholders, a move that Anders says shocked competitors but won praise in the Pentagon and on Wall Street. Anders's own earnings for 1992 put him in *Fortune* magazine's list of the top ten highest paid CEOs. In 1993 he stepped down as CEO in what appeared to be retirement, moving to an island off the coast of Washington State. In reality, he is still serving as General Dynamics' chairman of the board, still traveling, still wishing he had more time for boating.

Frank Borman, 1994

Sometimes, out on the water, he catches sight of the moon, and aside from a memory of a feeling of disbelief—he wonders if the whole thing was a very realistic simulation—he is wryly aware that he doesn't see it the way most others do.

"Other people will say, 'Oh, what a beautiful moon.' I'll think to myself, If they could just see it up close. It appears beautiful because it's a long ways off, and you can't see it's really a very uninviting place." Anders concedes that he would probably feel differently if he'd been able to see the moon the way his twelve moonwalking colleagues had. It is still one of Anders's regrets: for all his satisfaction at having flown Apollo 8, he would rather have been the last man to walk on the moon than the first to go around it. ☾

For many years Bill Anders noted the progress of the other astronauts as they tackled Life After the Moon, but he paid special attention to Frank Borman.

Borman had left NASA with international fame, but had been determined not to coast on his celebrity. His plate was full with lucrative offers to join anything from aerospace giants to broadcasting companies. A few wanted him for PR purposes, to be the "company astronaut," and that was something he wanted to avoid. In the end he went to work for Eastern Airlines. The airline business, he felt, was something he could sink his teeth into. After seven years as a senior vice president Borman was named Eastern's top executive in 1976. His gruff management style—along with his aversion to the lush corporate life-style of his predecessors—ruffled more than a few feathers among his executives. But soon, after a long decline in the 1960s, Eastern was enjoying the most profitable years in its history. Then, in 1978, deregulation hit the

After seven years as a senior vice president Borman was named Eastern's top executive in 1976.

airline industry, and Borman's troubles began to build. Beset by high labor costs and low employee morale, he found himself in a bitter power struggle with leaders of Eastern's unions that dragged on into the 1980s. The last traumatic confrontation came in early 1986, when Eastern's board accepted Frank Lorenzo's offer to buy the airline. In Borman's own words, it was his first failed mission. Says Anders, "I think he had a business problem that nobody could solve."

Meanwhile, Anders had been wishing for "some kind of Apollo 8 fraternity," and for a reunion with his former commander. For several years he'd marked the anniversary of the flight by sending his crewmates a telegram: LET'S GET TOGETHER. "After a while," Anders says, "I realized I wasn't even hearing back from Borman. . . . I finally gave up." Anders couldn't have been more surprised when, one day in the summer of 1986, his secretary said, "Frank Borman is on the line." When Anders picked up the phone—"Bill? Frank."—Borman sounded as if he'd talked to him the day before. The call turned out to be for advice on an airplane engine Borman was buying, but about a year later Borman called again, just to say hello. "If you ever come out here," he told Anders, "love to see you."

"That phone call was the first time that Frank Borman and I have communicated when he's not under pressure," Anders said a few weeks later. He speculated that Borman, who had moved to Las Cruces, New Mexico, where his son ran a car dealership, had gotten over the sting of his defeat at Eastern. Said Anders, "He's probably mellowed."

In 1988 the Apollo 8 fraternity finally convened at the Aerospace Hall of Fame in San Diego, in time for the flight's twentieth anniversary. Seeing his crewmates together for the first time in sixteen years, Anders had a chance to renew the friendships—and settle an old score. For years, Frank Borman had maintained that he had taken the famous picture of the first earthrise; he even said he'd had to grab the camera away from Anders to do it. A few years later, just for fun, Jim Lovell threw his hat in the ring, insisting *he* took the picture. To settle the issue, Anders had a NASA photography expert research the matter, and the results seemed to be in Anders's favor. But that didn't stop Borman from sticking by his story. Fortunately for Anders, Apollo 8's on-board voice recorder issued the final, unequivocal verdict: Anders took the picture. In San Diego, during a slide presentation before a group of high school students, up came the earthrise photo. ☾

"I'd better not comment on this one," Anders said. "Jim, why don't you comment?"

Lovell announced, "This picture has always been under contention . . . I think it's time now for Frank to have a public admission." Throughout the event, the earthrise picture became a running joke, and Borman took it all with good humor. Anders's suspicions about Borman were correct: he had indeed mellowed.

"I'm one of the few guys you may meet that's at peace with himself," Borman says. "First of all, I'm lucky to be alive. I made it through some very traumatic times. I've got the same healthy, supportive wife I started out with—a real accomplishment in the NASA group. And I've got two boys that we have good relationships with." And Las Cruces, he says, is a place where no one cares that he was president of Eastern, or that he went to the moon. ☾

At the reunion one evening, Borman entered a restaurant with his wife and some friends, and the maître d' stared at him for a long moment. Finally he said, "Didn't you used to be . . ."

"I used to be with an airline," Borman said. The young man didn't get the hint. Borman offered him another clue. "I used to be with NASA." Still no recognition. One of Borman's friends decided to end the suspense: "He used to be an astronaut." At last the maître d' realized he was meeting a man who had been to the moon. But before he could say anything Borman had the last word: "But now I'm just an old fart."

Alan Shepard, 1986

Years ago, Alan Shepard is said to have quipped, "Before I went to the moon, I was a rotten s.o.b. Now I'm just an s.o.b." To those who knew him, that statement alone—the fact that Shepard admitted any change at all—spoke volumes.

"I suppose I'm a little nicer now than I used to be," Shepard said in 1988. That, he added, had everything to do with getting to prove himself on Apollo 14. "Maybe it's because I wanted it so much, it was so important to me, and I did it, and I was able to relax and be a little more human."

It was Alan Shepard, ego still very much intact, who showed a surprising degree of balance when it came to looking back. Of his space experience, he said, "You put it in a box, put it on the shelf, put a ribbon around it, and move on to something else." But he did not avoid opening the box, for example, when the moon was up. "I never look at the moon anymore," Shepard said, "without thinking about it, just saying, '*Wow.*' Only that. Then going on and doing whatever you were doing."

In person, he seemed younger than his seventy years: tall, tanned, and fit, dressed casually but tastefully, he wore a gold chain around his neck bearing an Apollo 14 emblem. His brown hair, swept across his forehead, showed little gray. Alan Shepard was aging gracefully. And part of his maturing, he said, was the understanding that "there are things in life other than flying airplanes."

In the late 1980s Shepard served as chairman of a consulting business called 7-14 Enterprises. He was on the board of directors of twelve or fifteen companies, including Kmart and Kwik Copy. And he was president of the Mercury 7 Foundation, created by the Original 7 to give scholarships to college students studying space and engineering. None of the moon voyagers save Neil Armstrong had as much name recognition, and Shepard was able to use his fame to raise money for the foundation. Though he was more often identified as the first American in space, he was also the first lunar golfer, and participated in lucrative celebrity tournaments. And it was for the foundation that Shepard was happy to look back.

"I don't mind going back and opening up the box again from time to time," Shepard said of his speaking engagements. "And I find it satisfying for me to do it and make some money for this foundation. To be able to share that experience with someone is fun, and at the same time create moneys for youngsters in college that want to get involved in scientific pursuits, I find that very satisfying."

A splendid moon, ripening toward full, graces the October dusk. As darkness settles on the town of Acton, Massachusetts, there is activity at the high school, and an air of expectancy. Cars stream into the parking lot, parents and children file into the large auditorium. This night they will hear the experiences of a man who walked on the moon. Even now, well before the presentation begins, he is seated in the front row, greeting some of his audience, signing autographs. Tall, white-haired, dark-suited, Gene Cernan has the larger-than-life look of a celebrity. That is something that has been reinforced by television; if you were watching in the weeks after the *Challenger* disaster, you probably saw him on "World News Tonight," or conversing with Ted Koppel from a giant "Nightline" screen. For more than twenty years Cernan has been making public appearances, and he still looks the way people expect an astronaut to look.

Cernan's audience this evening spans two generations. Many of the parents were in college when Cernan walked on the moon; most of them would not, if asked, have been able to identify him as the last to leave his footprints in the lunar dust. Nor would many of their children, who have grown up in an age in which space travel seems routine. For them, Apollo is a story in their history books. Tonight Cernan will talk to the parents, about the importance of education, about the need for America to find a goal in space to mobilize its technology and capture the imagination of its people. But first, he looks back. "I can take you to the surface of the moon so that you're standing next to me," Cernan likes to say, and standing on the stage he launches into a symphony of recollection. He tells of looking at the beauty of the sunlit earth, "surrounded by the blackest black you can conceive of. I can't show it to you, but I can tell you it exists, because I saw it with my own eyes." And

Gene Cernan, 1994

the audience is transported by the words of a space traveler.

One afternoon the previous July, in his Houston office, Cernan talked about his life, his goals, and his identity as a moonwalker. Among the memorabilia on the walls are pictures of Cernan at Taurus-Littrow, standing before sun-drenched mountains and the American flag. Cernan has never felt the need to put his past behind him; in fact, he seems to have integrated it into his identity. And when he came back from the moon, Cernan had no intention of living in the past. After Apollo 17 he was assigned to other projects within NASA and then, in 1976, decided to strike out on his own. There was no letdown, he says: "You get ready, and you charge, and it's like having another mission." Over the years Cernan has been involved in a variety of business ventures: serving as vice president of an energy company in Houston, helping to start up a small airline in St. Louis, working as a marketing consultant for a Boston-based high-technology firm. In 1981 he created the Cernan Corporation, which channels his energies into aerospace consulting and other activities, including helping to create space exhibits for museums around the world. Like many of his former colleagues, he maintains a breakneck schedule, gone for weeks at a time, often home for only a day or two on the weekend. Like them, he shows no signs of slowing down. But even now, Cernan confesses, he has been unable to find a pursuit that really satisfies him. He says, "It's tough to find an encore."

But it isn't at all difficult to find people who still want to hear about what Cernan did more than two decades ago. And he is happy to answer. When Cernan talks about his moon experiences, not a hint of boredom dulls his words, which sometimes tumble so fast upon one another that he seems to speak without punctuation. And the people Cernan wants to reach most are children. All too aware how crucial education is to our nation's future, he says, Cernan takes every opportunity to use his own space experiences as inspiration. He tells students, "I urge you to dream—I did, and one day I found myself standing on the surface of the moon." ☾

At the Acton high school, the children listen quietly while Cernan speaks and shows slides, but when he is done speaking to their parents, they have their own inquiries to make. A young girl asks, "How far can you jump on the moon?" Cernan lights up. "That's a good question. Does anyone know what gravity is?" Cernan had shown some nervousness before, talking to the adults, but it is gone now. For these children he describes the wonders of zero g, and how he opened a can of peaches and turned it upside down while he ate. He tells them mischievously, "If you want to learn to be a spaceman, go home tonight and try it." And there are more questions—How did you feel at liftoff? What color was the sky? Is it hot or cold in space? And Cernan answers them all, speaking well past the time anticipated, until a little girl asks the old standby: "What does it feel like when you're on the moon?"

Cernan smiles and bows his head for a moment. He begins to piece together an explanation. "You can move around very easily in one-sixth gravity, so it feels very comfortable. You're not warm, because your suit is air-conditioned with water." Then he interrupts himself and tries a different approach, one from the heart.

"I'll tell you what it feels like. It feels like you're dreaming. You wonder when you're going to wake up. It's almost like your mom told you a wondeful story when you went to bed and, you know, sugar plums—it's like Santa Claus has already come. Being on the moon is like Santa Claus just gave you your wish."

Gene Cernan likes to point out a pair of photographs that share a frame on his office wall: one of himself at Taurus-Littrow, the other of Neil Armstrong at Tranquillity Base. The last man to walk on the moon and the first. They could not be more different in the way they have handled their positions; just

Neil Armstrong, 1989

as Cernan has embraced public attention, Armstrong has removed himself from it. Armstrong avoids public appearances and turns down most requests for interviews; he says little of his public life and nothing at all about his private life. And at most astronaut reunions, Armstrong is notable by his absence. In truth, Armstrong's response to the position that fate has given him was entirely in character. ☾

In the fall of 1969, after the Apollo 11 postflight world tour, Armstrong returned to the Astronaut Office, but not for very long. He wanted to fly in space again, he says, "but when they asked me to move up to Washington, I guess that indicated to me that they had other thoughts of what I ought to be doing than flying." That turned out to be running NASA's aeronautics activities, a post that Armstrong held for a year. He left for one reason: he wanted to teach. In 1962 he had mentioned to the other astronauts that he planned to write an engineering textbook some day. No one was surprised when, in September 1971, Armstrong became an engineering professor at the University of Cincinnati.

In doing so, Armstrong left the mainstream behind and returned to the solitude of his native Ohio. He and his wife, Jan, purchased a dairy farm near the small city of Lebanon. To the rest of the world it seemed, as Mike Collins wrote in his memoir *Carrying the Fire,* that by leaving Washington, Armstrong had retreated to his castle and pulled up the drawbridge. At the mention of this, Armstrong laughs and says, "You know, those of us who live out in the hinterlands think that people that live *inside* the Beltway are the ones that have the problems."

"You know, those of us who live out in the hinterlands think that people that live inside the Beltway are the ones that have the problems."

—Neil Armstrong

Robert Hotz, formerly the editor of *Aviation Week* magazine and a longtime friend of Armstrong's, himself a farmer in rural Maryland, says he completely understands what lured Armstrong away from NASA.

"Hell, you're in this high-tension world of aerospace. You get out on the farm. You look at the mountains across the valley, which are several million years old and are going to be there through the life of the planet. You understand that you're a short-term phenomenon, like the mosquitoes that come in the spring and fall. You get a perspective on yourself. You're getting back to the fundamentals of the planet. Neil feels that way, because we've talked about it, and so do I."

If Armstrong has kept a low profile, he has hardly been a recluse. In 1979 he left teaching for a variety of business activities. In 1986, he served as vice chairman of the presidential commission to investigate the *Challenger* disaster. He has appeared in advertisements for Chrysler, and has even hosted a cable TV documentary series on the history of flight. He gives assistance to historians who research Apollo. The common thread, in each case, is that Armstrong has met the world on his own terms, engaging only in activities he judges favorably. In short, Armstrong has handled the demands of his fame by rationing himself. While some of his colleagues, and others at NASA, wish he were a more visible spokesman for the agency and for the cause of space exploration, most of the others praise his approach. They say Armstrong was the ideal "choice" for the role of first man on the moon, as if it were an office to be filled. Were he more visible, they say, he would cheapen its currency.

Armstrong says he understands that fame is a direct result of media exposure, and that he has received a disproportionate share of it. Privately, he has said he can't understand why everyone focuses on the first lunar landing more than the other flights; after all, Apollo was a group effort. To him the title First Man on the Moon has little meaning, since in his mind the landing itself was the flight's most significant accomplishment—a feat he and Buzz Aldrin achieved at the same instant. But in general, the public and the media do not see it his way. Even other astronauts describe Armstrong's position as unique. However much they are besieged by phone calls and letters—from autograph seekers, space enthusiasts, documentary filmmakers, corporate executives—Armstrong gets more of it. One moonwalker says that when it comes to fame, if the astronauts were football players they would be a high school team, and Armstrong would be "the only guy in the NFL."

Perhaps the most famous picture of an astronaut is a photograph of Buzz Aldrin on the moon. He is facing the camera, standing at the edge of a small crater; in his mirrored visor is a tiny image of the photographer, Neil Armstrong, and the Sea of Tranquillity. If Armstrong's journey since that photograph was taken has been one of the most private, then Aldrin's has been among the most public and, at times, the most difficult.

Aldrin's troubles began shortly after he emerged from quarantine and began months of public appearances. Just as he had anticipated, he was uncomfortable with the spotlight and the role of PR spokesman for his agency. In 1971, after a period working on the Space Shuttle program, he decided to leave NASA to resume his air force career. Aldrin had hoped to be named commandant of cadets at the Air Force Academy; instead, after a ten-year absence from active duty, he was thrust into command of the test pilot school at Edwards. Aldrin, a non-test pilot with no managerial experience, found himself unable to perform up to his own high standards; his self-esteem was eroded. Within a year he was hospitalized for depression. Aldrin, who had spent his life striving for almost superhuman achievement, faced a devastating reality: he was sick and he needed help.

Two years later, he described his struggles against manic-depression and alcoholism in a confessional autobiography called *Return to Earth.* By going public, and in the process, debunking the myth of the perfect astronaut, Aldrin won himself both praise and criticism. But Aldrin was determined to tell the world what he went through, and to let others with similar problems know that they have company, even among men who walked on the moon. In doing so, he communicated his own humanity.

In the book, Aldrin charted a number of factors, including his own relentless drive for achievement, spurred by an intense and driven father. There had been other cases of depression in his family. Today Aldrin believes that given his family environment, coupled with his own personality, and possibly a genetic predisposition to the disease, he would probably have been destined for alcoholism late in life even if he had not gone to the moon. Instead, the crisis was accelerated by the double jeopardy that many returning moon voyagers faced: the sudden onslaught of world attention, followed almost immediately by the end of their astronaut careers. Aldrin, more than most of his colleagues, felt the loss: When Apollo 11 ended, so did his sense of purpose; that only made the hero's mantle harder to wear. And when he retired from the air force in 1972, he lost the ordered life he had thrived on since 1947, when he entered West Point. Without that structure, Aldrin says, his struggle only became more difficult. As he told an interviewer in 1993, "There I was—introverted, supersensitive, a perfectionist, concerned about what people thought of me. Jesus, it was a setup. No wonder I was in trouble." ☾

Buzz Aldrin, 1995

By 1984 Aldrin's recovery was long completed. He had gone six years without a drink; his depression was under control. But somehow, a new purpose eluded him. The years at MIT and NASA, when he helped pioneer techniques for space rendezvous, were still a bright memory. Sometimes, looking into the shaving mirror, he told himself he would probably never experience that kind of creativity again.

But Aldrin was wrong. That year he again found inspiration in the intricate rhythms of orbital mechanics. This time, his innovation was the cycler, a spacecraft that, if placed on the proper trajectory, would circle continuously

John Young, 1983

between the earth and Mars, ferrying astronauts and cargo with little need for fuel. Soon he was hard at work developing the cycler concept, along with ideas for an earth-orbit spaceport. By returning to the activity that gave him the most happiness—using his creative powers to advance space exploration—Aldrin had rediscovered his calling.

Today, happily remarried, trim and energetic, Aldrin is a one-man think tank, designing everything from new launch vehicles to scenarios for returning to the moon. But unlike thirty years ago, when Apollo was in full swing, he has found himself confronting a space program, its future clouded by uncertainty, that has not been ready for his innovations. Aldrin is undaunted.

He spends his days networking with engineers, space advocates, and others who share his vision. As chairman of the National Space Society, he works to generate public support for future space activities. It isn't an easy way to make a living; Aldrin admits that he is probably too outspoken for most aerospace consulting work. "I don't want to tell the client what he wants to hear," he explained in 1992. "I want to tell them what I think." Instead of a livelihood, Aldrin says, his work is "an expensive hobby." As much as possible, he offsets his expenses by making personal appearances. And if his life still harbors frustrations and pitfalls—Aldrin says he has to be careful not to overwhelm himself with too many projects—then it has also given him great satisfaction. Today, with the struggles of the 1970s long behind him, he talks about his life with surprising openness and self-awareness. "I'm lucky," Aldrin says. "*I'm alive.* And . . . I'm a better person for having gone through that. I got a chance to redo my life."

For John Young, the man whom Lee Silver calls "the archetypical extraterrestrial," whether to remain an astronaut was never an issue. During NASA's long spaceflight hiatus of the 1970s, when his Apollo colleagues left to pursue other challenges, Young stayed where he was. By 1978 only a handful of veterans remained at the Johnson Space Center, where the astronaut corps was now infused with a new breed of young pilots, physicians, and scientists, men—and, for the first time, women—who had grown up with the reality of space travel. Some had been in grade school when Young copiloted the first Gemini flight. Now he was their chief, having taken over the office in 1974. At the time, the move had surprised a number of his colleagues who wondered how someone as quirky and difficult to know as Young would fare as a manager. No one questioned his dedication, his drive, or his engineering prowess. And he was still on flight status, training to command NASA's most important mission since the first moon landing: the maiden voyage of the reusable Space Shuttle.

On the day the shuttle program was approved by Congress back in 1972, Young was on the moon, and when he heard the news he said excitedly, "The country needs that shuttle mighty bad. You'll see." But making the shuttle a reality under the project's extremely tight budget taxed NASA's ingenuity. The shuttle orbiter would be a space plane the size of a DC-9, capable of flying 100 missions. To save money, designers were forced to scrap plans for a reusable, liquid-fueled booster. Instead the orbiter, mated to a fuel tank as big as a grain silo, would use its own liquid-fueled engines, with help from two powerful—and risky—solid rocket boosters. In all, the shuttle would be the most sophisticated flying machine ever built. Young focused his energies on the monumental technical difficulties of getting it ready for flight.

In April 1981, three years later than originally planned, the shuttle *Columbia* finally soared into orbit. When the TV camera was turned on, there was Young in the commander's seat, reading glasses on the bridge of his nose, smiling and saying, "The vehicle's performing like a champ." They had called Al Shepard the "Old Man" when he flew at age forty-seven; Young was fifty. A day later, after a reentry at twenty-five times the speed of sound, Young steered *Columbia* to a perfect landing on a desert runway at Edwards. Later, after the ground crews had arrived, Young emerged and bounded down the stairway to inspect his ship, punching the air with his fist like a relief pitcher who had just won the World Series. That day, Young told a crowd of well-wishers, "We're really not too far, the human race isn't, from going to the stars."

But John Young wasn't about to leave NASA. He told an interviewer, "I live the space program. I breathe it. I eat it. I sleep it. . . ."

For a time it seemed that John Young's spaceflight career might be just as limitless. Late in 1983 he was back in orbit, this time as commander of a nine-day science mission called Spacelab 1. By January 1986, with twenty-four successful shuttle flights behind them, NASA was planning its most spectacular year since Apollo, with fifteen new launches. Young was to be part of it; he was unofficially slated for his seventh trip into space, this time to deploy NASA's scientific showpiece, the $1.5 billion Hubble Space Telescope. Friends say that he was as excited as a kid about the assignment. Then the shuttle *Challenger* exploded seventy-three seconds after launch, killing its seven-member crew, including schoolteacher Christa McAuliffe. The explosion shattered public perceptions about the seemingly routine nature of spaceflight, and about NASA's infallibility. Within the agency, amid shock and bewilderment, came the sudden realization that they had all been living too close to the edge. Young had bad dreams about the disaster for weeks.

In the aftermath, Young became an outspoken critic of his agency. One of his memos, citing an "awesome" list of potentially serious safety problems with the shuttle, was leaked to the press. NASA managers had repeatedly risked lives, it said, by letting the pressure of heavy launch schedules override safety concerns. And the agency was in danger of another accident with its plans for a fast-paced, post-*Challenger* launch schedule. Some colleagues said later that he was trying, in the only way he knew how, to protect his people. Feeling responsible, Young had strongly wanted to remain as chief astronaut until the shuttle was flying again. But in April 1987, in his twenty-fifth year as an astronaut, he was suddenly taken out of the Astronaut Office and made special assistant for engineering, operations, and safety to center director Aaron Cohen. Some, including Young, suspected that his criticisms had prompted the move, an idea that Cohen publicly denied. But the main reason seemed to be that it was simply time for him to move on. "I don't think *anybody* should stay in the same job for ten years," said one former high-level NASA official. "The problem is, you create legends bigger than life. I don't think it's good to create legends."

But John Young wasn't about to leave NASA. He told an interviewer, "I live the space program. I breathe it. I eat it. I sleep it. . . . I'm not willing to give it up as long as I can make a contribution to it." No one ever said Young's spaceflight career was over, and he continued to hope for another shuttle mission. And while he said he doesn't hold his two lunar voyages above any of his other flights—"You can't compare 'em"—Young maintained an enduring interest in the moon.

When the geologists held their annual Lunar and Planetary Science Conference, Young was always there, keeping up with new developments. In the mid-1980s he heard of the theory that still provides the best explanation for the origin of the moon: Soon after the earth formed, it was struck by an asteroid the size of Mars. So violent was this cataclysm that it vaporized parts of the earth's mantle and ejected the vapor into space, where it formed a disk of material that eventually coalesced to form the moon. ☾

But at the Johnson Space Center, the moon had long faded from view. Apollo had receded into a nostalgic past, along with the ample space budgets of the 1960s. But in the early eighties a handful of space center scientists began to study scenarios for going back to the moon. Around the center their ideas were often greeted by skepticism or even ridicule, but John Young was one of their best allies, always encouraging, always interested. Meanwhile, outside NASA, interest was building in the far more audacious goal of sending humans to Mars. For these advocates the Red Planet was what the moon had been in 1961: mysterious, just beyond reach, awaiting the footsteps of human beings. To them, Mars was the goal NASA sorely needed to revitalize the space program and excite the American public. A return to the moon, they said, would only siphon off necessary resources for the Mars trip. The moon, they said, is boring.

"It's anything but boring," Young said quietly during an interview at the Johnson Space Center in 1989. "We don't even begin to understand it." When Young talks about the moon he speaks from the experience of confronting its unknowns, not only at the Descartes highlands but from an orbiting command module. He reminds us that it is a world with one-quarter the land area of the earth. "To think that twelve guys went there and we've figured it out, that's crazy."

Young also knows of the moon's enormous potential for astronomers. A telescope on the lunar surface would be free of the turbulence and light pollution that mars the view through the earth's atmosphere. All wavelengths of light would be accessible to it. On the lunar far side, radio telescopes would be free of any man-made interference. Telescopes of unprecedented size and resolving power could be built in the moon's one-sixth g. The late Harlan Smith of the University of Texas talked of building giant "nirvana telescopes" powerful enough to study the type in a newspaper on earth—or to see surface features in planets orbiting other stars.

Even more compelling, the moon may provide an answer to the planet's pressing energy needs. Houston-based engineer David Criswell has designed solar power stations that could be set up on the moon and used to relay energy, in the form of microwaves, to receiving stations on earth. Manufactured from lunar materials, the first stations could produce 50 billion watts per year, Criswell says. Within forty years, if more were built, they could supply a whopping 20 trillion watts—the energy budget of the entire planet. The cost would be enormous, perhaps just shy of a trillion dollars. But selling power to the earth at a modest ten cents per kilowatt-hour, the venture could turn a profit after five years of operation. ☾

Meanwhile, a group of scientists researching nuclear fusion have zeroed in on the moon's supply of helium-3, which was discovered in the Apollo soil samples in the early seventies. This isotope of helium is exceedingly rare on earth, but it is literally all over the moon; it is one of an assortment of gases deposited by the solar wind, the steady stream of charged particles emanating from the sun. A fusion reactor that used helium-3 as fuel, say the scientists, would be cleaner, safer, and more efficient than the systems now being developed. And since a metric ton of the gas could provide as much energy as about $3 billion worth of coal, it would be worth going to the moon to get it. Mining helium-3 would be a simple matter of scooping up moon dust, heat-ing it, and collecting the gases that are driven off. And for the occupants of a lunar base, the process would also yield huge amounts of water, hydrogen, methane, and other gases valuable for survival and industry. In time, the scientists claim, helium-3 mining could sustain not only the earth's energy needs, but the life of an industrial community on the moon. It could even provide fuel for the Mars voyagers. Young says simply, "I think we should go put a base up there. That's what I think."

By many accounts, NASA was only a shadow of the agency that had gone to the moon twenty years earlier.

But in July 1989, as the twentieth anniversary of the first lunar landing approached, that seemed far from likely. By many accounts, NASA was only a shadow of the agency that had gone to the moon twenty years earlier. Young understood that a number of factors had contributed to the agency's decline, including a lack of direction from the top. What was necessary, he said, was for the president to give NASA, and the country, a new mission in space, to do what Kennedy had done twenty-eight years before.

On July 20, 1989, George Bush proposed the Space Exploration Initiative, a thirty-year effort that would include a permanent manned space station, a lunar base, and manned expeditions to Mars. And Young was full of enthusiasm. "I'd like to see it as a crash program," he said. He thought people would put their lives into it, the way they did in the old days.

By the end of 1992, the Space Exploration Initiative was dead, though it never officially disappeared from NASA's agenda. Two years in a row, the start-up funds had been voted down by Congress. In Houston, NASA's Office of Exploration, created to plan long-range programs, was disbanded. Meanwhile, space station *Freedom* was so mired in bureaucracy that Deke Slayton called it an "aerospace WPA." While outsiders questioned the station's value, NASA rewrote its purpose and reworked its design again and again. With $8 billion already spent, there was little more to show than a set of blueprints. By the following year, newly elected president Bill Clinton told NASA to cut costs and redesign the station yet again, then declared that it would become a joint pro-

Stu Roosa, 1994

gram with the Russians. For a time in 1993, with the space station's fate still uncertain, the space center in Houston had become a place where, in the words of one thirty-year veteran, "people . . . are just holding down their jobs." By year's end, shuttle astronauts had successfully carried out a difficult repair mission in earth orbit to save the ailing Hubble Space Telescope. And for the first time in a long while, NASA's future looked brighter.

On a trip through Egypt in 1976, Stu Roosa and his wife visited a granite quarry near Aswan, where they saw an unfinished obelisk, perhaps thirty-five hundred years old. Had it been completed, this 1,100-ton monolith would have stood 137 feet high, the largest of its kind. But sometime before the artisans had finished their work, the stone cracked and they abandoned it where it is today, partially emerged from a rock slab on the valley floor. When Roosa thought about Apollo, he often remembered the sight of that quarry in Aswan.

"I always thought Apollo was our unfinished obelisk," said Roosa. "It's like we started building this beautiful thing and then we quit." He shook his head with a mixture of sadness and disbelief. "History will not be kind to us, because we were *stupid.*"

Today, at the NASA space centers in Houston and Florida, the Saturn Vs for Apollo 18 and 19 lie on tourist stands, like unfinished obelisks, reminders of a time that seems now as remote as the moon itself. Across the distance of a quarter century, Apollo is an anomaly. There was a rare confluence of historical forces in 1961: A perceived threat to national prestige from the Soviet Union was met by a dynamic leader, John Kennedy, and economic prosperity allowed him to launch a massive effort to demonstrate America's capabilities. The moon was the ideal target—close enough to reach, audacious enough to capture the imagination.

Apollo happened so quickly that it all seems unreal. Eight years after Kennedy spoke to Congress, his challenge was met. "We couldn't *think* about it in that length of time today," says Ken Mattingly. "I tell all my friends, We could not go to the moon today. We *can not* do it."

Mattingly speaks out of frustration. For several years, as a consultant on the space station project, he saw firsthand the bureaucracy, the resistance to new ideas, and most important, the lack of national will. Of the station he says, "The damn country needs it but they don't know why!" Mastering the technology necessary to build a permanent workplace in earth orbit, he says, is essential preparation for tomorrow's far voyages.

"If you're ever going to Mars, you need that space station." Today, with the Apollo corporate memory nearly depleted from NASA and industry, Mattingly warns that we are in danger of forgetting how to explore space. "If you don't build things, you don't know *how* to build things. We can't handle a ten-year hiatus. There won't be anybody left. So is it a WPA? Maybe so, but it's a WPA with a purpose."

Today Mattingly works for General Dynamics' aerospace division in San Diego. It doesn't bother him that the nation lost interest in Apollo; that, he says, is a mark of progress—you have to accept the last step before you can take the next one. And if the glory days are behind us, Mattingly says, then that's where they should be. "Apollo was the adolescence of space. Now we have to grow up and be adults." To Mattingly, that means the unglamorous but crucial work of making space pay its way, and he is trying to do just that. At General Dynamics Mattingly is working on revamping the venerable Atlas booster, a descendent of the rocket that put John Glenn in orbit in 1962. He hopes the Atlas will help open up cheaper access to space, so that the next generation can do the things he can only dream about. And when space begins to turn a profit, Mattingly says, then it will be time to explore again.

Ken Mattingly, 1982

"We *will* go to Mars. And who knows what we'll find? Once again it will be the journey that is the true test, as much as what you learn when you get there." Mars comes no closer than 35 million miles; that trip, Mattingly says, is *really* leaving home. With present technology, the trip there will take six months. And don't expect the Mars voyagers to get the kind of attention that Mattingly's colleagues got: "When we send people to Mars, by the time they get there, and they call back, people are going to say, *Who?*" If Apollo was an anomaly, then the astronaut as national hero was equally so. And Mattingly stresses that he and the other moon voyagers were just symbols of the entire enterprise. They received the most attention, he says, but they didn't make the greatest contribution. Don't put astronauts on a pedestal: "There are extraordinary people *everywhere* in life . . . who are just as competent, just as cool, just as anything else you can mention. . . . All they need is an opportunity." Mattingly doesn't dwell on being one of the handful of men to go to the moon, though he says, "I wouldn't *trade* that for anything." ☾

In Mattingly's mind what stands out most is what happened to him not in space but on earth. "It was being part of a team that was dedicated to something that transcended individual aspirations. That's what Apollo was. It was thousands of people who were willing to work day and night. . . . You can't imagine what that's like compared to an everyday experience."

It doesn't take space exploration to bring out the best in us, he says. If there is a lesson to be learned from Apollo, it is that we can do difficult things, when the objective is clearly defined, and when enough people and funds are dedicated to accomplishing it. From 1961 to 1972, the objective was as clear and inspiring as any you could ask for. You had only to go outside at night and look at it.

The indirect purpose . . . of all music played on the terraces of the audiences of the moon, seems to be to produce an agreement with reality.

—Wallace Stevens, *The Necessary Angel*

Even as he circled the moon, Alan Bean had promised himself to live his life the way he wanted to. In a sense, his vow was the opposite of Pete Conrad's. Bean was happy to let his moon trip influence him; he says it gave him the courage to drastically change his life. Today, the walls of Bean's condominium in suburban Houston are covered with mementos of his two spaceflights. But the main room of the house, the one where Bean spends most mornings, is the painting studio. Displayed around his easel are images Bean has created: Neil Armstrong, his gold visor reflecting the Sea of Tranquillity, unfurling

Alan Bean, 1994

an American flag in the vacuum; Ed Mitchell in midstride, map in hand, on his way to Cone crater; and Bean himself, poised in the footpad of the lunar module *Intrepid,* about to take his first step onto the lunar surface. Alan Bean has left spaceflight to record, as only he can, the experiences of the first human beings to visit another world. Nothing else, he says, could have lured him away from the Astronaut Office.

Bean did not leave abruptly. In July 1973, a month after Pete Conrad returned from Skylab, Bean took off for the station with his own crew. He didn't come home until the start of autumn. In their fifty-nine days in orbit Bean and his crewmates zapped the productivity meter so far off the scale that it was only halfway through the *next* Skylab mission that mission control stopped thinking there was something wrong with the third crew, whose performance was closer to normal. For Bean, like Pete Conrad, Skylab stands out above his moon flight—but not for the same reason. Circling the earth for weeks on end, Bean says, was less risky than going to the moon—but more demanding:

"I had a better personal feeling at the end of Skylab than I did at Apollo 12, *because* it took more self-discipline to work doing similar things day after day after day. On the moon mission it was different every day, and, I mean, anybody could do well on that, anybody that had been trained." In the end Bean is most proud of himself not for bravery, but productivity.

Then came the long wait for the Space Shuttle. Bean was getting more and more serious about painting, a hobby he had begun in 1962, in night classes while at Pax River. When the new crop of astronauts arrived Bean was in charge of training them, but occasionally he would take a couple of weeks off to try out the life of an artist. It could not have been more different from the life he'd led as a naval aviator and an astronaut, not only in substance but in image: jet pilots don't paint. But the more he painted, the more he realized he could do credit to the Apollo program in art. More than a century ago Frederic Remington captured the spirit of the Old West with his paintings of cowboys. In our time, Bean realized, one of the quintessential images is the

Bean has likened himself to a kind of tribal storyteller; his goal is to capture the moments that were important, or special, to the other astronauts.

moonwalker. In 1981, a few months after the first shuttle flight, Bean announced he was leaving NASA to become a full-time artist. His colleagues asked, "Are you sure you can make a living at that?" Bean wasn't sure, but he wasn't going to let that stop him. "If I can go to the moon," he told himself, "I can learn to be a good artist."

That happened with the aid of several Houston art teachers, especially a wildlife painter who became his mentor. The more Bean learned about painting, the more he realized that it shares an analytical quality with flying. Where he was once alert to the subtle change in pitch of a jet engine or the idiosyncrasies of an Apollo fuel cell, Bean now zeros in on the play of reflected light within a shadow, the subtle rainbow hues that hide on seemingly bland objects. Bean has brought his characteristic attention to detail to his new line of work. He is an expert on every variation of space suit, lunar module, and lunar roving vehicle. Near the easel sit the gloves Bean wore on space walks during his Skylab mission; now he uses them as a reference. It may take

Bean more than a month to finish a painting; he is aiming for two hundred finished works over twenty years.

Bean has likened himself to a kind of tribal storyteller; his goal is to capture the moments that were important, or special, to the other astronauts. Sometimes he calls the astronaut he is painting to ask a question about the event. Sometimes, the others tell him their suggestions for new paintings. Bean's paintings have been exhibited in Houston galleries and have sold to private collectors and corporations. But of the customers Bean really wishes he could attract—the other astronauts—only one, shuttle astronaut Claude Nicollier, has bought one. He wryly admits they may not see $15,000 worth of art in one of his paintings. He wishes they did; he is sure they would enjoy owning their own lunar portrait more than they may realize. He seems wistful, saying this; he still sees in his colleagues the same loner tendencies that kept the Astronaut Office from feeling like a squadron. He says he wishes they could all network with each other and help each other; at the same time he knows that is against their nature.

Bean has thought about his fellow moon voyagers and the paths they have taken, and he believes there is a common thread. "I think that everyone who went to the moon came back more like they already were." And on the whole, his theory seems to hold up. Jim Irwin was a religious man before he went. Ed Mitchell was already interested in psychic phenomena. Neil Armstrong's retreat from the media spotlight, and Gene Cernan's acceptance of it, are entirely consistent with the people they were before Apollo. From Borman to Schmitt, Bean says, going to the moon only magnified tendencies that were already present.

And for Bean himself, going to the moon gave him the courage to choose a very different life. He did not know, when he made that promise to himself in lunar orbit, that it would lead to the painting studio. But looking back, he says, he was just following the call that was already inside him, and that, to quote the poet Robert Frost, has made all the difference. "I've been really happy since the moon trip," Bean says. "I feel blessed."

When Bean looks at the moon, he searches for the small bright dot that marks the place, out in the Ocean of Storms, where he and Pete Conrad walked. He wishes there would be some mystical feeling, but there never seems to be. He says he would love to go back, to see the moon with his new eyes, to find the subtle rainbows hiding in the gray-tan plains. If he could do it again, he'd put it all on hold for five minutes—the experiments, the sample gathering, the reporting—and just take it all in. *This is five minutes for me . . .*

●●◐○○○○◑●

Project Apollo remains the last great act this country has undertaken out of a sense of optimism, of looking forward to the future. That it came to fruition amid the upheaval of the sixties, alongside the carnage of the Vietnam War, only heightens the sense of irony and nostalgia, looking back twenty-five years later. By the time Apollo 11 landed, we were already a changed people; by the time of Apollo 17, we were irrevocably different from the nation we had been in 1961. It is the sense of purpose we felt then that seems as distant now as the moon itself. If NASA has lost direction, it is only because we have not chosen to give it one. Instead of letting the moon be the gateway to our future, we have let it become a brief chapter in our history. The irony is that in turning away from space exploration—whose progress is intimately linked to the future of mankind—we rob ourselves of the long-term vision we desperately need. Any society, if it is to flourish instead of merely survive, must strive to transcend its own limits. It is still as Kennedy said: Exploration, by virtue of difficulty, causes us to focus our abilities and make them better.

It is left to a future generation to return, to go about their work in the light of the earth, to pick up where Apollo left off. For a time it will still be magical to meet someone who has been to the moon, but gradually that mystique will fade, and a moon voyager will seem no more extraordinary than an explorer who has been to the Antarctic. And the moon will seem, as Jim Lovell likes to say, a little bit closer, only three days away.

But for the rest of this century, and probably for the rest of their lives, the men who have been to the moon still possess that special uniqueness that they are so quick to disavow. Until someone follows in their still-preserved footsteps, we are left to make the journey in our imaginations. It is a journey we can make, and should make, with our children: to look at the moon, and see Tranquillity, where Armstrong and Aldrin walked upon a stage one midsummer night, and the Ocean of Storms, and Fra Mauro, and Hadley, and Descartes, and Taurus-Littrow. To trace the paths of the orbiting command modules. To preserve not just the facts, the events, but the wonder.

In the end, when we confront the fact that human beings have been to the moon, we continue to have magic thoughts about the experience. We hold the magic, the awe, not the moon voyagers themselves. They flash quickly upon it and move on. The significant journey takes place not in their minds, but in ours.

APOLLO LANDING SITES

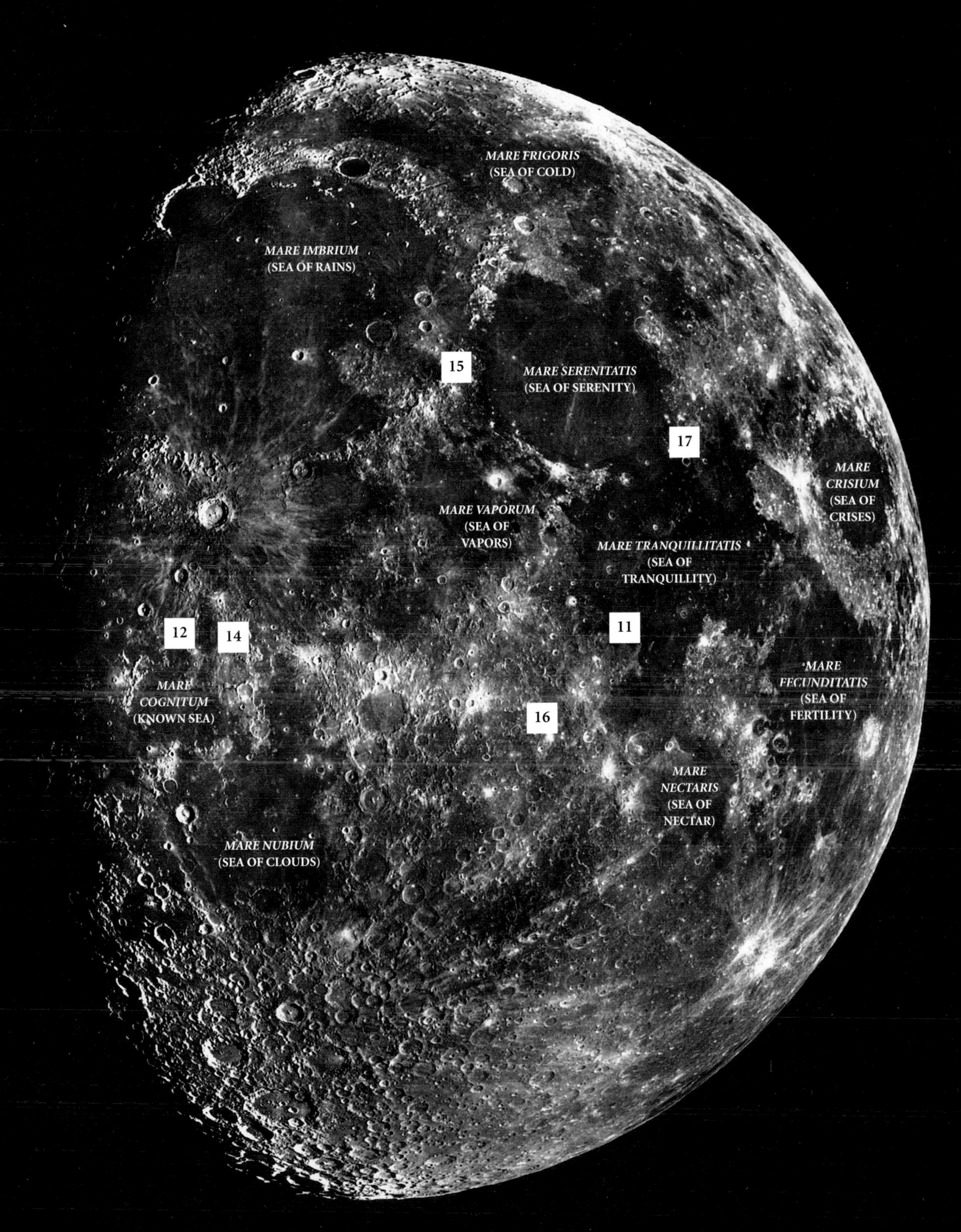

APPENDIX A:

ASTRONAUT BIOGRAPHICAL INFORMATION

(source: Hawthorne, *Men and Women of Space*)

EDWIN EUGENE ALDRIN, JR. (legally changed to Buzz Aldrin, 1979)
born: January 20, 1930, Montclair, New Jersey
education: B.S., United States Military Academy, 1951;
Sc.D., Massachusetts Institute of Technology, 1963
marriage: Joan A. Archer, 1954 (divorced); Beverly Van Zile, 1975 (divorced);
Lois Driggs, 1987
children: James, 1955; Janice, 1957; Andrew, 1958
spaceflights: Gemini 12, Apollo 11
activities: Aldrin lives in Laguna Beach, California,
where he develops ideas for future space transportation systems.

WILLIAM ALISON ANDERS
born: October 17, 1933, Hong Kong; considers La Mesa, California, to be his hometown
education: B.S., United States Naval Academy, 1955;
M.S. (nuclear engineering), Air Force Institute of Technology, 1962;
completed Harvard Business School Advanced Management Program, 1979
marriage: Valerie Hoard, 1955
children: Alan, 1957; Glen, 1958; Gayle, 1960; Gregory, 1962; Eric, 1964; Diana, 1972
spaceflights: Apollo 8
activities: Anders lives on an island off the coast of Washington state.
He serves as chairman of the board of General Dynamics.

NEIL ALDEN ARMSTRONG
born: August 5, 1930, Wapakoneta, Ohio
education: B.S. (aeronautical engineering), Purdue University, 1955
marriage: Janet Shearon, 1956 (divorced); remarried, 1994
children: Eric, 1957; Karen, 1959 (deceased); Mark, 1963
spaceflights: Gemini 8, Apollo 11
activities: Armstrong lives on a farm in Lebanon, Ohio, and serves as
chairman of AIL Systems, Inc., a defense electronics firm based in Deer Park, New York.
He is also on the boards of several corporations.

ALAN LAVERN BEAN
born: March 15, 1932, Wheeler, Texas; considers Forth Worth, Texas, to be
his hometown
education: B.S. (aeronautical engineering), University of Texas at Austin, 1955;
graduated from Navy Test Pilot School, Patuxent River, Maryland, 1960, and the School of
Aviation Safety, University of Southern California, 1962
marriage: Sue Ragsdale, 1955 (divorced); Leslie Gombold, 1982
children: Clay, 1955; Amy, 1963
spaceflights: Apollo 12, Skylab 2
activities: Bean lives in Houston, Texas, where he is a full-time artist.

FRANK BORMAN
born: March 14, 1928, Gary, Indiana; grew up in Tucson, Arizona
education: B.S., United States Military Academy, 1950; M.S. (aeronautical engineering),
California Institute of Technology, 1957; graduated from the Air Force Experimental Flight
Test Pilot School, Edwards Air Force Base, California, 1960; graduated from the Air Force
Aerospace Research Pilot School, 1961; completed Harvard Business School Advanced
Management Program, 1970
marriage: Susan Bugbee, 1950
children: Frederick, 1951; Edwin, 1953
spaceflights: Gemini 7, Apollo 8
activities: Borman lives in Las Cruces, New Mexico. He is chairman and CEO of the California-

based Patlex Corp., which manages patents for laser technology. He also serves on the boards of several corporations.

EUGENE ANDREW CERNAN

born: March 14, 1934, Chicago, Illinois
education: B.S. (electrical engineering), Purdue University, 1956; M.S. (aeronautical engineering), Naval Postgraduate School, Monterey, California, 1964
marriage: Barbara Atchley, 1961 (divorced); Jan Nanna, 1987
children: Teresa ("Tracy"), 1963
spaceflights: Gemini 9, Apollo 10, Apollo 17
activities: Cernan is president of the Cernan Corporation, an aerospace management and marketing consulting firm in Houston, Texas.

MICHAEL COLLINS

born: October 31, 1930, Rome, Italy; considers Washington, D.C., to be his hometown
education: B.S. (military science), United States Military Academy, 1952; graduated from Aircraft Maintenance Officer School, 1958; graduated from Squadron Officer School, 1959; graduated from the Air Force Experimental Flight Test Pilot School, 1961; graduated from the Air Force Aerospace Research Pilot School, 1963; graduated from the Industrial College of the Armed Forces (by correspondence), 1971; attended Harvard Business School Advanced Management Program, 1974
marriage: Patricia Finnegan, 1957
children: Kathleen, 1959; Ann, 1961; Michael, 1963 (deceased)
spaceflights: Gemini 10, Apollo 11
activities: The author of several books, including *Carrying the Fire: An Astronaut's Journeys* and *Flying to the Moon and Other Strange Places,* Collins has retired to North Carolina, where he says he spends more time fishing than he does writing.

CHARLES CONRAD, JR.

born: June 2, 1930, Philadelphia, Pennsylvania
education: B.S. (aeronautical engineering), Princeton University, 1953; graduated from Naval Test Pilot School, 1958
marriage: Jane DuBose, 1953 (divorced); Nancy Fortner, 1990
children: Peter, 1954; Thomas, 1957; Andrew, 1959; Christopher, 1960 (deceased)
spaceflights: Gemini 5, Gemini 11, Apollo 12, Skylab 1
activities: For 20 years beginning in 1976, Conrad served as a vice president of the McDonnell Douglas Corporation. In 1996 Conrad formed Universal Space Lines, whose goal is to provide commercial launch services much as airlines operate today.
died: July 8, 1999, Ojai, California, from injuries sustained in a motorcycle accident

CHARLES MOSS DUKE, JR.

born: October 3, 1935, Charlotte, North Carolina; considers Lancaster, South Carolina, to be his hometown
education: B.S. (naval sciences), United States Naval Academy, 1957; M.S. (aeronautics and astronautics), Massachusetts Institute of Technology, 1964; graduated from the Air Force Aerospace Research Pilot School, 1965
marriage: Dorothy Claiborne, 1963
children: Charles, 1965; Thomas, 1967
spaceflights: Apollo 16
activities: Duke, who lives in New Braunfels, Texas, has retired from business to pursue a Christian ministry. His experiences as a born-again Christian, he says, are so much more fulfilling than his astronaut career that he calls walking on the moon "the dust of my life" in comparison.

RONALD ELLWIN EVANS

born: November 10, 1933, St. Francis, Kansas
education: B.S. (electrical engineering), University of Kansas, 1956; M.S. (aeronautical engineering), Naval Postgraduate School, 1964
marriage: Janet Pollom, 1957
children: Jaime, 1959; Jon, 1961
spaceflights: Apollo 17
activities: Evans lived with his wife, Jan, in Scottsdale, Arizona, where he was involved in a

number of private business ventures, including real estate. He was also active in public speaking, to convey his spaceflight experiences.
died: April 7, 1990, Scottsdale, Arizona, of a heart attack

RICHARD FRANCIS GORDON, JR.

born: October 5, 1929, Seattle, Washington
education: B.S. (chemistry), University of Washington, 1951; graduated from the Naval Test Pilot School, 1957; graduate student in operations analysis at the Naval Postgraduate School, 1963
marriage: Barbara Field, 1953 (divorced); Linda Saunders, 1981
children: Carleen, 1954; Richard, 1955; Lawrence, 1957; Thomas, 1959; James, 1960 (deceased); Diane, 1961
spaceflights: Gemini 11, Apollo 12
activities: Gordon, who lives in Los Angeles, California, is president of Space Age America, a consulting firm for developing space museums and theme parks.

FRED WALLACE HAISE, JR.

born: November 14, 1933, Biloxi, Mississippi
education: B.S. (aerospace engineering), University of Oklahoma, 1959; graduated from the Air Force Aerospace Research Pilot School, 1964
marriage: Mary Grant, 1954 (divorced); F. Patt Price, 1979
children: Mary, 1956; Frederick, 1958; Stephen, 1961; Thomas, 1970
spaceflights: Apollo 13
activities: Haise is president of Grumman Technical Services, Inc., in Titusville, Florida, which handles government contracts for engineering, computer operations and maintenance, and aircraft maintenance.

JAMES BENSON IRWIN

born: March 17, 1930, Pittsburgh, Pennsylvania
education: B.S. (naval science), United States Naval Academy, 1951; M.S. (aeronautical engineering), M.S. (instrumentation engineering), University of Michigan, both 1957; graduated from Squadron Officer School; graduated from the Air Command and Staff College; graduated from the Air Force Experimental Flight Test Pilot School, 1961; graduated from the Air Force Aerospace Research Pilot School, 1963
marriage: Mary Monroe, 1959
children: Joy, 1959; Jill, 1961; James, 1963; Jan, 1964; Joe, 1969 (adopted by the Irwins in 1973)
spaceflights: Apollo 15
activities: Irwin left NASA in 1972 to form High Flight, a Christian ministry.
died: August 8, 1991, Glenwood Springs, Colorado, of a heart attack

JAMES ARTHUR LOVELL, JR.

born: March 25, 1928, Cleveland, Ohio; considers Milwaukee, Wisconsin, to be his hometown
education: B.S., United States Naval Academy, 1952; graduated from the Naval Test Pilot School, 1958; graduated from the School of Aviation Safety, University of Southern California, 1961; completed Harvard Business School Advanced Management Program, 1971
marriage: Marilyn Gerlach, 1952
children: Barbara, 1953; James, 1955; Susan, 1958; Jeffrey, 1966
spaceflights: Gemini 7, Gemini 12, Apollo 8, Apollo 13
activities: Lovell lives in Horseshoe Bay, Texas; he devotes his energies to public speaking and other activities to disseminate information on the space program.

THOMAS KENNETH MATTINGLY II

born: March 17, 1936, Chicago, Illinois
education: B.S. (aeronautical engineering), Auburn University, 1958; graduated from Air Force Aerospace Research Pilot School, 1966
marriage: Elizabeth Dailey, 1970 (separated; deceased, 1991)
children: Thomas, 1972
spaceflights: Apollo 16; Space Shuttle missions STS-4 and STS 51-C
activities: Mattingly heads the Atlas booster program for General Dynamics in San Diego, California.

EDGAR DEAN MITCHELL

born: September 17, 1930, Hereford, Texas; considers Artesia, New Mexico, to be his hometown
education: B.S. (industrial management), Carnegie Institute of Technology (now Carnegie-Mellon University), 1952; B.S. (aeronautical engineering), Naval Postgraduate School, 1961; Sc.D. (aeronautics and astronautics), Massachusetts Institute of Technology, 1964; graduated from Air Force Aerospace Research Pilot School, 1966
marriage: Louise Randall, 1951 (divorced); Anita Rettig, 1973 (divorced); Sheilah Ledbetter, 1989 (divorced)
children: Karlyn, 1953; Elizabeth, 1959; Adam, 1984; Mitchell also adopted his second wife's children by a previous marriage: Paul, 1963, and Mary, 1964
spaceflights: Apollo 14
activities: Mitchell is a freelance management consultant in Boca Raton, Florida.

STUART ALLEN ROOSA

born: August 16, 1933, Durango, Colorado; grew up in Claremore, Oklahoma
education: B.S. (aeronautical engineering), University of Colorado, Boulder, 1960; graduated from the Air Command and Staff College; graduated from the Air Force Aerospace Research Pilot School, 1965; completed Harvard Business School Advanced Management Program, 1973
marriage: Joan Barrett, 1957
children: Christopher, 1959; John, 1961; Stuart, 1962; Rosemary, 1963
spaceflights: Apollo 14
activities: Roosa was president of Gulf Coast Coors in Gulfport, Mississippi.
died: December 12, 1994, Falls Church, Virginia, of complications from pancreatitis

HARRISON HAGAN SCHMITT

born: July 3, 1935, Santa Rita, New Mexico; grew up in Silver City, New Mexico
education: B.S. (geology), California Institute of Technology, 1957; studied at the University of Oslo, Norway, under a Fulbright Fellowship, 1957-58; Ph.D. (geology), Harvard University, 1964
marriage: Theresa Fitzgibbon, 1985
children: none
spaceflights: Apollo 17
activities: A former U.S. senator, Schmitt is now an independent consultant in science, technology, and public policy in Albuquerque, New Mexico.

DAVID RANDOLPH SCOTT

born: June 6, 1932, Randolph Air Force Base, Texas
education: B.S. (military science), United States Military Academy, 1954; M.S. (aeronautics and astronautics), Massachusetts Institute of Technology, 1962; graduated from the Air Force Aerospace Research Pilot School, 1963
marriage: Lurton Ott, 1959
children: Tracy, 1961; Douglas, 1963
spaceflights: Gemini 8, Apollo 9, Apollo 15
activities: Scott, who lives in Manhattan Beach, California, is president of Scott Science and Technology, a consulting firm for developing space transportation and exploration concepts.

ALAN BARTLETT SHEPARD, JR.

born: November 18, 1923, East Derry, New Hampshire
education: B.S., United States Naval Academy, 1944; graduated from the Naval Test Pilot School, 1951; graduated from the Naval War College, Newport, Rhode Island, 1958
marriage: Louise Brewer, 1945
children: Laura, 1947; Juliana, 1951. The Shepards also raised a niece, Alice.
spaceflights: (Mercury-Redstone) MR-3, Apollo 14
activities: Shepard was president of the Mercury Seven Foundation, which raises scholarship funds for college students in science and engineering.
died: July 21, 1998, Pebble Beach, California, of leukemia

THOMAS PATTEN STAFFORD

born: September 17, 1930, Weatherford, Oklahoma
education: B.S., United States Naval Academy, 1952; graduated from the Air Force Experimental Flight Test Pilot School, 1959
marriage: Faye Shoemaker, 1953 (divorced); Linda Dishman, 1988
children: Dionne, 1954; Karin, 1957
spaceflights: Gemini 6, Gemini 9, Apollo 10, Apollo-Soyuz Test Project (joint Soviet-American flight in 1975)
activities: Stafford is vice chairman of Stafford, Burke and Hecker, a consulting firm in Alexandria, Virginia. He is also a consultant with Defense Technology in Oklahoma City.

JOHN LEONARD SWIGERT, JR.

born: August 30, 1931, Denver, Colorado
education: B.S. (mechanical engineering), University of Colorado, 1953; M.S. (aerospace science), Rensselaer Polytechnic Institute, 1965; M.B.A. (business administration), University of Hartford, 1967
marriage: never married
children: none
spaceflights: Apollo 13
activities: Swigert took leave from NASA in 1973 to serve as executive director of the House of Representatives Committee on Science and Technology in Washington, D.C. In 1977, he left that post and formally resigned from NASA. That year he made an unsuccessful bid for a U.S. Senate seat. In 1982 he announced his candidacy for representative from Colorado's newly created 6th Congressional District. During the campaign, Swigert underwent surgery to remove a malignant tumor from his right nasal passage. He continued to campaign vigorously, but the cancer spread to his bone marrow. On November 2, 1982, he won the House seat. By December 19 the cancer had spread to his lungs, and Swigert was hospitalized. He died a week later, just eight days before his Congressional term would have begun.
died: December 27, 1982, Georgetown University Hospital, Washington, D.C., of respiratory failure resulting from lung cancer

ALFRED MERRIL WORDEN

born: February 7, 1932, Jackson, Michigan
education: B.S. (military science), United States Military Academy, 1955; M.S. (aeronautical and astronautical engineering), M.S. (instrumentation engineering), University of Michigan, 1963; graduated from the Instrument Pilots Instructor School, Randolph Air Force Base, Texas, 1963; graduated from the Empire Test Pilots School, Farnborough, England, 1965; graduated from the Air Force Aerospace Research Pilot School, 1965
marriage: Pamela Vander Beek, 1955 (divorced); Sandra Wilder, 1974 (divorced); Jill Hotchkiss, 1982
children: Merrill, 1958; Alison, 1960
spaceflights: Apollo 15
activities: Worden handles technology acquisition and business development for B.F. Goodrich Aerospace in Brecksville, Ohio.

JOHN WATTS YOUNG

born: September 24, 1930, San Francisco, California; considers Orlando, Florida, to be his hometown
education: B.S. (aeronautical engineering), Georgia Institute of Technology, 1952; graduated from the Naval Test Pilot School, 1959
marriage: Barbara White, 1955 (divorced); Susy Feldman, 1972
children: Sandy, 1957; John, 1959
spaceflights: Gemini 3, Gemini 10, Apollo 10, Apollo 16, (Space Shuttle) STS-1, STS-9/Spacelab 1
activities: Young is special assistant to the director of the Johnson Space Center for engineering, operations, and safety.

APPENDIX B:

PERSONS INTERVIEWED

ASTRONAUTS

Buzz Aldrin; Joseph P. Allen; William A. Anders; Neil A. Armstrong; Alan L. Bean; Frank Borman; Gerald P. Carr; Eugene A. Cernan; Michael Collins; Charles Conrad, Jr.; R. Walter Cunningham; Charles M. Duke, Jr.; Anthony W. England; Ronald E. Evans; Edward G. Gibson; Richard F. Gordon, Jr.; Fred W. Haise, Jr.; Karl G. Henize; James B. Irwin; Joseph P. Kerwin; James A. Lovell, Jr.; T. Kenneth Mattingly II; James A. McDivitt; Edgar D. Mitchell; Robert A. Parker; Stuart A. Roosa; Walter M. Schirra, Jr.; Harrison H. Schmitt; Russell L. Schweickart; David R. Scott; Alan B. Shepard, Jr.; Donald K. Slayton; Thomas P. Stafford; Alfred M. Worden; John W. Young

ASTRONAUT WIVES

Joan Aldrin; Valerie Anders; Susan Borman; JoAnn Carr; Jane Conrad; Dotty Duke; Jan Evans; Marilyn Lovell; Joan Roosa; Lurton Scott; Beth Williams

ASTRONAUT CHILDREN

Jaime Evans; Tracy Scott

ASTRONAUT OFFICE MANAGER, KENNEDY SPACE CENTER

Charles Friedlander

APOLLO PROGRAM MANAGERS, MANNED SPACECRAFT CENTER

George Abbey; Owen Morris

APOLLO PROGRAM MANAGERS, NASA HEADQUARTERS

Rocco Petrone; Samuel Phillips

CREW SYSTEMS ENGINEERS

David Ballard; Michael Brzezinski; John Covington; Raymond Zedekar

EXPERIMENT SCIENTISTS

Marcus Langseth; Jack Trombka

FLIGHT CONTROLLERS

John Aaron; Steven Bales; Ronald Berry; Charles Deiterich; Robert Legler; John Llewellyn; Edward Pavelka

FLIGHT DIRECTORS

Clifford Charlesworth; Peter Frank; Gerald Griffin; Eugene Kranz; Glynn Lunney

GEOLOGISTS

Robin Brett; Uel Clanton; John Dietrich; Michael Duke; Farouk El-Baz; Clifford Frondell; James Head; Friederich Hörz; Elbert King; Robert Laughon; Gary Lofgren; Harold Masursky; William Muehlberger; William Phinney; Gerald Schaber; Eugene Shoemaker; Leon Silver; Gordon Swann; George Ulrich; Edward Wolfe

HISTORIANS

Roger Bilstein; David Compton; Richard Hallion; Eric Jones; John Logsdon; Alex Roland

JOURNALISTS

Howard Benedict; Robert Hotz; Roy Neal; Wayne Warga

MISSION PLANNERS

Rodney Rose; John Sevier; Howard Tindall

NASA ADMINISTRATORS

Thomas Paine; James Beggs

NASA AIRCRAFT OPERATIONS
Joseph Algranti

NASA MANAGERS, HEADQUARTERS
Chester Lee; Gerald Mossinghoff; Willis Shapley

NASA MANAGERS, MANNED SPACECRAFT CENTER
Maxime Faget; Christopher Kraft

NASA PUBLIC AFFAIRS OFFICERS
Jack Riley; Julian Scheer; Douglas Ward; Terry White

NASA X-15 AND B-52 PILOTS
Johnnie Armstrong; William Dana; Milton Thompson

NAVY SQUADRON PILOTS, VA44
Jack Keating; Jack Raider

NORTH AMERICAN AVIATION
Joe Cuzzupoli; Charles Feltz; George Jeffs

SIMULATOR INSTRUCTORS AND SUPERVISORS
Pleddy Baker; David Bragdon; Charles Floyd; Jay Honeycutt; Frank Hughes; Robert Pearson; Carl Shelley; David Strunk; Michael Wash

SPACECRAFT ENGINEERS
Charles Mars

SPACE SUIT TECHNICIANS
Al Rochford; Troy Stewart

UNITED STATES INFORMATION AGENCY
Simon Bourgin

APPENDIX C:

APOLLO MISSION DATA

APOLLO 7 OCTOBER 11-22, 1968
crew: Walter M. Schirra, Donn F. Eisele, R. Walter Cunningham
description: First manned earth-orbit test of the Apollo command and service modules.
mission duration: 10 days, 20 hours, 9 minutes

APOLLO 8 DECEMBER 21-27, 1968
crew: Frank Borman, James A. Lovell, Jr., William A. Anders
description: First manned flight around the moon. Apollo 8 orbited the moon 10 times on Christmas Eve, 1968.
time in lunar orbit: 20 hours, 7 minutes
mission duration: 6 days, 3 hours, 1 minute

APOLLO 9 MARCH 3-13, 1969
crew: James A. McDivitt, David R. Scott, Russell L. Schweickart
description: Earth-orbit test of the entire Apollo spacecraft. Included rendezvous maneuvers between the command module and lunar module, and a 38-minute space walk by Schweickart to test the lunar space suit and backpack.
spacecraft (command module, lunar module): *Gumdrop, Spider*

mission duration: 10 days, 1 hour, 1 minute

APOLLO 10 MAY 18-26, 1969

crew: Thomas P. Stafford, John W. Young, Eugene A. Cernan
description: Dress rehearsal for the lunar landing.
spacecraft (command module, lunar module): Charlie Brown, Snoopy
time in lunar orbit: 2 days, 13 hours, 41 minutes
mission duration: 8 days, 0 hours, 3 minutes

APOLLO 11 JULY 16-24, 1969

crew: Neil A. Armstrong, Michael Collins, Edwin E. Aldrin, Jr.
description: First lunar landing.
spacecraft (command module, lunar module): *Columbia, Eagle*
time in lunar orbit: 2 days, 11 hours, 34 minutes
lunar landing date, location: July 20, Sea of Tranquillity
time on lunar surface: 21 hours, 36 minutes
moonwalk duration: 2 hours, 31 minutes
pounds of samples collected: 47.7
mission duration: 8 days, 3 hours, 18 minutes

APOLLO 12 NOVEMBER 14-24, 1969

crew: Charles Conrad, Jr., Richard F. Gordon, Jr., Alan L. Bean
description: Second lunar landing. Conrad and Bean were first to make a pinpoint lunar landing, touching down some 600 feet from the unmanned Surveyor 3 probe.
spacecraft (command module, lunar module): *Yankee Clipper, Intrepid*
time in lunar orbit: 3 days, 17 hours, 2 minutes
lunar landing date, location: November 19, Ocean of Storms
time on lunar surface: 1 day, 7 hours, 31 minutes
moonwalk durations: 1st: 3 hours, 56 minutes
2nd: 3 hours, 49 minutes
pounds of samples collected: 75.7
mission duration: 10 days, 4 hours, 36 minutes

APOLLO 13 APRIL 11-17, 1970

crew: James A. Lovell, Jr., John L. Swigert, Jr., Fred W. Haise, Jr.
description: Third lunar landing attempt; aborted following the explosion of an oxygen tank inside the service module.
spacecraft (command module, lunar module): *Odyssey, Aquarius*
mission duration: 5 days, 22 hours, 54 minutes

APOLLO 14 JANUARY 31-FEBRUARY 9, 1971

crew: Alan B. Shepard, Jr., Stuart A. Roosa, Edgar D. Mitchell
description: Third lunar landing, and the first successful mission devoted entirely to scientific exploration of the moon.
spacecraft (command module, lunar module): *Kitty Hawk, Antares*
time in lunar orbit: 2 days, 18 hours, 40 minutes
lunar landing date, location: February 5, Fra Mauro
time on lunar surface: 1 day, 9 hours, 30 minutes
moonwalk durations: 1st: 4 hours, 47 minutes
2nd: 4 hours, 34 minutes
pounds of samples collected: 94.4
mission duration: 9 days, 0 hours, 2 minutes

APOLLO 15 JULY 26-AUGUST 7, 1971

crew: David R. Scott, Alfred M. Worden, James B. Irwin
description: Fourth lunar landing. First of the extended scientific expeditions, called J-missions, featuring extended lunar stay time, long-duration backpacks, and the battery-powered Lunar Rover. The Apollo 15 service module was the first to be equipped with the new Scientific Instrument Module. Worden performed a 38-minute space walk to retrieve scientific film during the trip back to earth.
spacecraft (command module, lunar module): *Endeavour, Falcon*
time in lunar orbit: 6 days, 1 hour, 17 minutes

lunar landing date, location: July 30, Hadley-Apennine
time on lunar surface: 2 days, 18 hours, 54 minutes
moonwalk durations: standup: 33 minutes
1st: 6 hours, 32 minutes
2nd: 7 hours, 12 minutes
3rd: 4 hours, 49 minutes
pounds of samples collected: 169
mission duration: 12 days, 7 hours, 12 minutes

APOLLO 16 APRIL 16-27, 1972

crew: John W. Young, T. Kenneth Mattingly II, Charles M. Duke, Jr.
description: Fifth lunar landing; first exploration of the moon's central highlands. Included a space walk by Mattingly lasting 1 hour, 24 minutes.
spacecraft (command module, lunar module): *Casper, Orion*
time in lunar orbit: 5 days, 5 hours, 53 minutes
lunar landing date, location: April 20, Descartes highlands
time on lunar surface: 2 days, 23 hours, 2 minutes
moonwalk durations: 1st: 7 hours, 11 minutes
2nd: 7 hours, 23 minutes
3rd: 5 hours, 40 minutes
pounds of samples collected: 208.3
mission duration: 11 days, 1 hour, 51 minutes

APOLLO 17 DECEMBER 7-19, 1972

crew: Eugene A. Cernan, Ronald E. Evans, Harrison H. Schmitt
description: Sixth and final Apollo lunar landing. Schmitt became the first professional scientist to land on the moon. First U.S. manned night launch. Longest Apollo flight. Space walk by Evans lasted 1 hour, 7 minutes.
spacecraft (command module, lunar module): *America, Challenger*
time in lunar orbit: 6 days, 3 hours, 48 minutes
lunar landing date, location: December 11, Taurus-Littrow
time on lunar surface: 3 days, 2 hours, 59 minutes
moonwalk durations: 1st: 7 hours, 11 minutes
2nd: 7 hours, 36 minutes
3rd: 7 hours, 15 minutes
pounds of samples collected: 243.1
mission duration: 12 days, 13 hours, 51 minutes

CHAPTER 9: THE SCIENTIST

14 Schmitt's awful puns: Two examples: One of the tools used by astronauts on the moon is called gnomon; it's used as a reference in photographs of lunar rocks to indicate local vertical. One of Schmitt's favorite phrases was, "Gnomon is an island."

When Schmitt encountered Paul Gast, the Manned Spacecraft Center's chief geologist, in the hall he would say loudly, "I'm appalled! I'm aghast!" much to the irritation of the high-strung Gast.

Gene Cernan turned the tables on Schmitt one day when the pair were training for Apollo 17. Driving a mockup of the Lunar Rover across a formation called an alluvial fan, Cernan hit a bump that dislodged Schmitt, who wasn't wearing his seat belt. Cernan looked over at Schmitt, lying on the ground, and said into his radio, "Well, the Schmitt just hit the fan."

15 *Shoemaker had made the first detailed geologic map of part of the moon:* Before Shoemaker's map, a less detailed geologic map of the entire moon had been prepared by Robert Hackman and Arnold Mason of the USGS in Washington. It is unclear how much they influenced Shoemaker's work, or vice versa.

17 *NASA rejected all three:* Schmitt, by all signs in perfect health, was rejected simply because he'd had surgery to correct a congenitally malrotated colon.

21 *most of the pilots had little or no enthusiasm for geology classes:* There were some exceptions. Gordon Swann says he saw a fair amount of interest from Wally Schirra, Scott Carpenter, Gordon Cooper, and several others among the Old Heads. He credits his late colleague Dale Jackson, one of the astronauts' early instructors, with this achievement. Some of the rookies, like Roger Chaffee, saw geology as their ticket to the lunar surface, and they worked hard at it; Chaffee was among the most promising students in the office.

21 *When Bill Anders wanted extra preparation:* More often than not, Schmitt's meetings with Anders took place in the sauna in the astronaut gym—not only because that was one of the few opportunities Anders could fit into his hectic schedule, but also, Schmitt suspected, to keep Borman from finding out.

23 *Armstrong turned in an excellent performance on the moon:* In particular, geologists have praised the variety of the rocks Armstrong collected during his hurried minutes on the moon.

24 *Schmitt called on his friends at Harvard and Caltech for help:* Shoemaker had become the chairman of Caltech's Division of Geological Sciences in January 1969.

24 *Silver made some pivotal refinements to the method:* Silver's method, which applies to rocks between 10 million and 4.5 billion years old, has become the standard means of dating the oldest rocks. The oldest rocks yet discovered on earth are between 3.8 and 4 billion years old.

24 *Silver, with his grad student Mike Duke:* Duke was one of the geologists who applied for the scientist-astronaut selection in 1965.

31 *Schmitt became a simulator hound:* The flight directors, whom Schmit had gotten to know better than he knew many of the astronauts, were happy to have him act as the astronaut for their own training runs, something the pilots rarely offered to do.

CHAPTER 10: A FIRE TO BE LIGHTED

39 *the first of the so-called J-missions:* At one time, NASA planned a series of I-missions, which were to involve extensive scientific observations from lunar orbit, without a landing, but these missions were never developed.

39 *a place called Hadley:* To the geologists, it was officially known as Hadley-Apennine.

39 *the lava plains that formed the valley floor:* Strictly speaking, these lava plains are not part of Mare Imbrium, but are instead part of an isolated patch of *mare* called Palus Putredinis, the Marsh of Decay.

42 *Jim Irwin . . . the only person who could have gone to the moon with him:* USGS geologist Jerry Schaber, who participated in briefings with Scott and Irwin, remembers that Scott tried to look after his lunar module pilot. After one hefty dinner at the crew quarters Lew Hartzell brought out a coconut cream pie. When Irwin reached for it Scott said, "Ah, Jim, you better watch it. Tomorrow's our [preflight] weigh-in." Irwin looked sad; Scott promptly took a helping of pie. He didn't have any problem keeping his weight down, Schaber said, but apparently Irwin did.

42 *this chunky white rock, called anorthosite:* On earth, anorthosite formed many times during the Precambrian. Not all of it is white; due to minute impurities it may be black or green. In rare cases—particularly in one locality in northern Labrador—it exhibits a spectacular play of color called Labradorescence.

46 *The geologists could hardly believe the resistance:* Gordon Swann says he had to fight to get a geologic hammer onto the missions. The engineers asked him why the astronauts needed a hammer to break rocks; couldn't they just pick up small ones? Swann explained that the big rocks might be important samples. Then the engineers told Swann they were afraid that if an astronaut hammered on a boulder, flying rock chips would damage his space suit. Swann pointed out that in all his years in the field, his own clothing had never been pierced by a flying rock chip. Tests using a space-suited test subject dispelled the worry. But in the end, Swann says, the only reason the geologists were able to get a hammer onto the Apollo 11 stowage list was that the astronauts would need it to hammer in core tubes. When it came time for Apollo 12, Swann says, the geologists assumed there wouldn't be any more problems with the hammer. But the engineers said, "You already flew that experiment. Why do you need to fly it again?" Bill Muehlberger got the same question when he wanted Apollo 16 astronauts John Young and Charlie Duke to take a 500mm lens to the moon—after all, it had already been done on Apollo 15.

49 *they depended on the astronauts' words:* Every so often the listeners would hear a description of a very strange "rock," and as they listened they would realize they were hearing a description of a beer can, or a cowpie.

CHAPTER 11: TO THE MOUNTAINS OF THE MOON

I: "Exploration at Its Greatest"

58 *"the* Falcon *is on the Plain at Hadley":* The Plain was a name chosen by Scott as a reference to the parade ground at West Point.

67 *followed by Irwin's grin-and-bear-it laughter:* After the flight, Irwin received a plaque from the engineers who built the Rover, bearing a pair of opaque welder's goggles and the inscription: "To be worn in case you are ever again riding the Rover on the moon with Dave Scott driving."

67 *he'd even driven over it in simulations:* A specially built Rover simulator, using the same video camera and terrain model used in the lunar module simulator, allowed the men to practice driving across Hadley.

71 *how long the rock had been sitting there:* Analyses showed that the rock had been there for 500 million years.

II: High Point

75 *after the last of a regular series of evening geology briefings at the Cape:* The evening briefings were Scott's idea. He was hungry for more preparation from the geologists, and he scheduled the briefings when the day of meetings and simulator runs was over, when they would be free from interruptions. The geologists would come to the crew quarters for one of Lew

Hartzell's hearty dinners, then they would go to work. Any idea pertaining to the geologic exploration of the Apollo 15 landing site was up for discussion. At one point, a young geologist named Jim Head, who worked with Farouk El-Baz at Bellcomm, was leading a discussion about Hadley Rille. Scott asked, "If Jim and I were able to drive down to the bottom of the rille, would that be useful?" Head could barely contain himself—the bottom might be teeming with clues to the rille's origin—and he answered, "Oh, *yeah*—Jesus, you could spend the rest of your *life* down there!" When everyone realized what he had unwittingly said, the room exploded in laughter.

80 *Allen combined a youthful enthusiasm with a keen scientific mind:* He was also an accomplished flier. In pilot training, Allen had proved himself by winning the highest awards for formation flying, instrument flying, and acrobatics.

87 *One of them, based on declassified spy-satellite technology:* The driving force behind getting this camera, called the Panoramic Camera, onto Apollo was a Survey geologist named Hal Masursky. Masursky's detailed, engaging lectures about the moon made him one of the astronauts' favorites.

89 *Either there was no life out there, or the cosmos must be teeming with it:* Of all Worden's experiences on the flight, the view of the universe he had during lunar night appears to have affected him the most. He says it led him to wonder whether it was possible that human beings are descendants of alien beings who visited earth in the distant past.

III: The Spirit of Galileo

93, 95 *"the lesson of the moon's antiquity and changelessness":* Wilhelms, *To a Rocky Moon: A Geologist's History of Lunar Exploration,* p. 280.

97 *He reached into a pocket on his suit and pulled out a falcon feather:* A friend of Scott's at the air force academy in Colorado Springs had arranged it; it had been shed by the falcon that was the academy's mascot.

97 *Centuries before, the story was told, Galileo Galilei:* Historians doubt that Galileo actually performed the experiment. Instead, they say, he may have solved the problem by sheer power of reasoning. However, he may indeed have dropped objects from the Leaning Tower of Pisa for a different purpose: to evaluate the effects of air resistance on falling bodies of various shapes and sizes.

99 *a small aluminum figure, a stylized representation of a fallen astronaut:* Scott had commissioned the sculpture from a Dutch artist named Paul van Hoeydonck, whom he met in New York.

IV: The Final Selection

103 *Scott pulled out a stack of first-day covers:* The U.S. Postal Service issued a new stamp during the flight of Apollo 15. Ten years had passed since Alan Shepard's Mercury flight, and the stamp commemorated a decade of achievement by Americans in space. On it, two astronauts rode a Rover across a bright moonscape. A special first-day cover, hand-canceled by Scott in a brief ceremony on the lunar surface, was brought back to earth for display. When Scott's crew arrived on the carrier *Okinawa,* they arranged to buy copies of the new stamp, and affixed them to their own envelopes, then had them canceled at the ship's post office (Justice Department internal memo, December 6, 1978).

105 *"One of the most brilliant missions in space science ever flown":* Compton, *Living and Working in Space: A History of Skylab,* p. 240.

107 *When Cernan crashed his helicopter in the Banana River during training:* About a week before the launch of Apollo 14, Cernan was flying a helicopter over the Banana River, enjoying the clear air and the smooth, mirrorlike water—so smooth, in fact, that he misjudged his altitude and crashed into the river. The chopper exploded in flames, and Cernan had to dive into the water to escape being burned to death. He arrived back at the crew quarters as Shepard was having breakfast. Shepard looked up, astonished to see Cernan standing there, soaking wet in his scorched flight suit. Cernan, who during training had joked about

hoping Shepard would break his leg so he could take his place, said, simply, "You win."

107 *In his mind, only one thing mattered:* Slayton said in 1989 that Engle's performance on the backup crew of Apollo 14 wasn't the issue in the Apollo 17 crew selection; neither was the pressure from the scientific community.

109 *"Yes, sir, I'll do the best job I can":* Schmitt's memory is unclear, but he thinks that the phone call was probably from James Fletcher, who had become the NASA administrator in 1971.

109 *explaining to his kids that he wasn't going to the moon: New York Times,* August 20, 1971. Engle's children were twelve-year-old Laurie and nine-year-old Jon.

CHAPTER 12: THE UNEXPECTED MOON

I: Luna Incognita

125 *a high point of 69 miles and a low point of 9 miles—the so-called descent orbit:* From Apollo 14 on, the astronauts used the SPS engine to place the joined command module and lunar module into the descent orbit, a maneuver previously performed by the lunar module alone, using its descent engine. This was done in order to save the LM's fuel for the Powered Descent, allowing the lander to carry more payload to the lunar surface.

129 *Young and Duke would have to take a wave-off:* Wave-off is a term used in aviation when a pilot is told by ground controllers to delay a landing attempt or to land elsewhere.

134 *"You're right; we* wouldn't *have let you land!":* While Mattingly recalls that McDivitt was serious, that seems extremely unlikely, given the amount of data available to him and others in Houston and at Downey. McDivitt does not remember the conversation but says he must have been pulling Mattingly's leg.

137 *NASA wanted the geologic objectives to be as carefully planned as any other aspect of the missions:* Gordon Swann recalls that NASA engineers were always asking him and the other geologists for a "voice plan"—in other words, they wanted to know what the astronauts would say as they collected rocks. Swann tried to explain to them that there wasn't any way to *know* what the astronauts would say, since no one could predict with certainty what they would find. If he could write up a voice plan, he told them, he wouldn't need to send the astronauts to the moon.

III: ". . . Or Wherever Geologists Go"

174 *the KC-135, a converted cargo plane:* Anyone who has worked in the KC-135 knows that it can be a difficult experience. After each period of zero g, the plane pulls out of its dive, and weightlessness is replaced by an onslaught of twice-normal gravity. For an astronaut in a space suit—especially one that might be less than adequately cooled—it was even more unpleasant. In a training session of 40 parabolas or more, it was all some of the pilots could do to keep from getting sick—not something they wanted to happen inside a pressure suit. No wonder the KC-135 was nicknamed the Vomit Comet.

176 *"science advances most when its predictions prove wrong":* Wilhelms, *To a Rocky Moon: A Geologist's History of Lunar Exploration,* p. 284.

CHAPTER 13: THE LAST MEN ON THE MOON

I: Sunrise at Midnight

187 *Born in Liberia, he'd been taken aboard a slave ship at the age of twelve:* See "A Conversation with the Nation's Oldest Citizen," by Grover Lewis, *Rolling Stone,* February 1, 1973, pp. 22-26.

190 *determined to see their wings clipped:* A number of Apollo astronauts say that some of the Original 7 alienated people in the space center with their arrogance. As one example, one of the Fourteen described what he called the "Schirra effect": In meetings with the engineers, Schirra would take a strong position on some aspect of spacecraft design, and no one could

talk him out of it, even when it was clear that he was wrong. Finally, he would say, "If you'd been there"—that is, in space—"you'd know." Then he would walk out. Another astronaut, from the Nine, says Alan Shepard's involvement with banks and other lucrative business ventures generated resentment among some NASA managers.

Such history aside, it is clear that the stamp affair angered many at NASA. The transcripts of the congressional hearings have never been published, but one NASA official who was there remembers that Connecticut senator Lowell Weicker verbally assaulted NASA managers. And according to one astronaut, when Chris Kraft came back to Houston from the hearings, he told a room full of astronauts, "The [astronauts] went in there and were treated like heroes. I went in there and I was totally humiliated. That is never going to happen to me again."

191 *The sequencer, aware of its own error:* What was the cause of the problem that caused a $450 million lunar mission, and the efforts of thousands of people, to grind temporarily to a halt? As discovered later by technicians, it was a single defective diode in a printed circuit card within the sequencer. *Apollo 17 Mission Report* (NASA internal publication), p. 14-1; Baker, *The History of Manned Spaceflight,* p. 438.

201 *Schmitt had first raised the idea late in the spring of 1970:* Schmitt's efforts in 1970 focused on the last four landing missions, Apollo 16-19. The geologists were already pushing for a landing at Tycho for Apollo 16. For Apollo 17, Schmitt was suggesting Mare Orientale, the giant impact basin that straddles the boundary between the near and far sides. For 18, he proposed a landing at the lunar north pole, where scientists hoped permanently shadowed craters might contain deposits of volatile elements, now in the form of ice, that had once escaped from the lunar interior. Then, Apollo 19 would go to Tsiolkovsky. The "lunar mafia"—Schmitt's name for his flight-controller friends—had tracked down some Tiros communications satellites, which TRW would have sold to NASA for about $80 million (details from the *Apollo 17 Lunar Surface Journal,* edited by Eric Jones, in press).

205 At 2 days, 17 hours into the flight, the Mission Elapsed Time clock was advanced 2 hours and 40 minutes, to make up for the delay in Apollo 17's launch. Mission Elapsed Times of events in Chapter 13 reflect this change.

205 Challenger *crept down the last 25 feet:* In truth, the word "crept" is only appropriate for the very last portion of the descent. At one point when *Challenger* was very close to the surface, Schmitt told Cernan his descent rate was "a little high"—mild words, but the slight intensity of Schmitt's voice hints at his concern. Today Schmitt says, with a laugh, "Gene had an aggressive streak in his flying."

205 *to make his own tracks in ancient dust:* Schmitt's first step on the moon was almost his downfall. When he stepped off the footpad, his boot came down on the sloping side of a boulder that was partly covered by beads of impact-created glass—like tiny ball bearings—and his leg went out from under him. Still gripping the ladder, he managed to avoid the fall.

210 *though only he could see it, he said,* "Look at the light": Only Cernan could see the glow, because he had the only window in the boost protective cover.

210 *Cernan . . . would talk about them for the rest of his life:* Schmitt points out that he has also been talking about those sights ever since Apollo 17. "I can't believe people are still interested," Schmitt says, "but they are."

II: Apollo at the Limit

217 *once inside—if a student is clever enough to get in:* According to Schmitt, the students always are.

219 *an optical illusion caused at certain angles of illumination:* Today, most lunar scientists believe that the grooves Scott and Irwin saw were indeed an illusion due to lighting effects.

220 *Anxious to save time, he went as fast as he dared:* On level ground, going flat out, the Rover hit 12 kilometers per hour, about half the speed of a world-class marathon runner.

221 *the planners were too conservative when they calculated walkback limits:* Based on Shepard

and Mitchell's journeys at Fra Mauro, it was assumed that two astronauts forced to return to the LM on foot would achieve a speed of about one and two-thirds miles per hour. Since no one had ever been forced to put these calculations to the test—fortunately—Schmitt had been unable to offer a better estimate.

222 *the geologists suspected the landslide:* Schmitt says that, in reality, a better term is avalanche, which refers to a slide that is fluid in character. In the case of the light mantle, Schmitt says, gases implanted in the soil by the solar wind would have allowed grains of dust to slide past one another. He believes the gases probably dissipated as the material descended, so that the avalanche became a landslide.

227 *Cernan and Schmitt headed for . . . a crater called Shorty:* Shorty had been named for a character in San Francisco author Richard Brautigan's novel *Trout Fishing in America.*

231 *now there was a very real possibility that Shorty was volcanic:* Another possibility, in Schmitt's mind, was that Shorty was an impact crater that had fractured the crust, allowing volcanic gases to escape to the surface.

231 *The ever-tightening circle of oxygen consumption and walkback limit:* As the J-missions progressed, the walkback limit was observed with less and less conservatism. Cernan and Schmitt were significantly closer to theirs at Shorty than any astronauts before them had been.

233 *hundreds of miles down in the lunar mantle:* Lee Silver has estimated that the magma that produced the fire fountains may have originated some 300 miles below the surface, over a quarter of the way to the moon's center.

III: Witnesses to the Earthrise

241 *getting within 9 miles was good enough:* Like the previous three crews, Evans and his crewmates had spent about a day in the descent orbit, with a low point of 50,000 feet.

241 *Farouk El-Baz's last protégé in lunar orbit:* Evans was told about the orange soil discovery, and the next time he was over the landing site he could see patches of orange, subtle but definitely there. Soon he was spotting it in other places, especially along the southwest border of Mare Serenitatis. Why hadn't he seen this earlier? After the mission he and El-Baz agreed: if you know what you are looking for, you find it.

243 *He did not know who had put it there:* Cernan has dropped hints that it may have been himself who put the sign on *Challenger*'s front landing leg. Schmitt, meanwhile, doesn't remember seeing it at all and says it probably never existed: it was just something Cernan decided to say over the radio, and Schmitt played along.

246 *the violence of the impacts that formed the lunar basins:* Apollo helped reveal that the time prior to about 4 billion years ago was an extremely violent one, both on the earth and moon. Schmitt suggests that the impacts of giant asteroids, which also pounded the young earth, may have played a role in the origin of life on our own planet. Perhaps, Schmitt says, chemical components of these asteroids, along with the tremendous energy of their impacts, helped create the conditions necessary for life to arise.

257 *He hated the words:* In 1988 Schmitt said, "I'm madder now than I was then. Nixon was right."

263 *with no guarantee of getting those people back:* Schmitt points out that in the strict sense, the risk for him and his crewmates circling the moon was no greater—and perhaps less—than for people in a submarine in the deep ocean. The greatest impact, he says, was psychological: "You're *no longer of the earth.*" He likens it to the first sailors who left sight of land. Like most of his other perceptions about his flight, this one came about in retrospect—at the time, he says, he was too busy to think about such things.

EPILOGUE: THE AUDIENCES OF THE MOON

281 *Roosa's words are ironic:* There is nothing unusual about the fact that Roosa did not know

that his own crewmate had had a life-changing experience on the flight. Most of the astronauts have had little idea of what their colleagues took away from their lunar voyages. In part, this is because astronauts rarely asked each other about any aspects of their flights besides the technical ones. Even when they did, it wasn't necessarily a rich exchange. Dave Scott remembers that during a party after Apollo 14, he asked Alan Shepard what it had been like on the moon. Shepard answered in one word: "*Spectacular.*" And that, Scott says, was all he needed to hear; he could see the rest in Shepard's eyes.

283 *there is almost nothing left of the real one:* According to NASA's lunar sample curator, Jim Gooding, about half the Genesis Rock is still kept inside the lunar sample vault in Houston. However, it has been broken up into small fragments by scientists searching for minerals that could be used to establish a more accurate age for the rock. But the Genesis Rock is an unusual example. Gooding points out that despite the large number of ongoing studies of lunar material—NASA sends out more than a thousand prepared samples to scientists each year—researchers are able to conduct their analyses using tiny amounts of rock or soil. As a result, most of the Apollo sample collection has not been touched. Some 74 percent by weight remains pristine within the Houston facility. Another 14 percent is kept in a special reserve storage facility in San Antonio. At least a tiny portion of every lunar sample has been studied, except for two core tubes collected on Apollo 16 and 17. For more information on the lunar sample facility, see the author's article, "Pieces of the Sky," *Sky & Telescope,* April 1982, pp. 344-49.

283 *By publicly admitting he had made a mistake, Irwin took himself off the pedestal:* In interviews with the author, Irwin confessed some resentment that the other astronauts did not come to the defense of him and his crewmates. Like Scott and Worden, he maintained that previous astronauts had done similar things, but that none would admit it. The author was not able to validate this claim in off-the-record conversations with other astronauts.

In 1983, Al Worden filed suit against NASA to return the first-day covers the agency had confiscated from him and his crewmates in 1972. In an out-of-court settlement, the Justice Department directed NASA to return the covers, and according to Worden, one official told him that the Apollo 15 crew had committed no wrongdoing by their actions regarding the first-day covers before and after their flight.

287 *he would rather have been the last man to walk on the moon:* Many astronauts have shared Anders's regret at not having walked on the moon. For Apollo 13's Fred Haise, the disappointment was compounded by the experience of a failed mission. But Haise's commander has a different attitude. If Jim Lovell could pick the flights he would like to be on, even with clear hindsight, one of them would be Apollo 13. It doesn't take anything extraordinary to do what is expected of you, Lovell says, but fighting for his life 200,000 miles from home tested him in ways that even a lunar landing wouldn't have. Says Lovell, "Apollo 13 was a test pilot's mission." He regrets that the Society of Experimental Test Pilots never recognized him and his crew for their performances. And when Congress awarded the Space Medal of Honor to Pete Conrad, Neil Armstrong, and a handful of other astronauts, Lovell was disappointed once more. A medal of honor, he points out, is given for action above and beyond the call of duty. Happily, Lovell was awarded the medal in 1995.

Apollo 13 has also had special meaning for Ken Mattingly. If people expect that he is glad for the twist of fate that kept him from the hardships Jim Lovell and his crew endured, glad to have a successful flight on his résumé instead of a failed one, then they are wrong.

"I have personally prospered by not being onboard," Mattingly says. "But I wish I had been. That's where I belonged. If it had been a success, if it had all worked out well, I wouldn't feel the way I do. But it meant more adversity than you would expect, and as a result I feel like somehow . . . I didn't do my share." He adds, "Does that make any sense? Hell, no."

289 *the famous picture of the first earthrise:* Here is Borman's earthrise story, as he told it to the author in March 1988:

"I'm looking over the lunar horizon, and there's the earth coming up. And I'm saying, 'Bill, take that picture! Get that one!' He says, 'I can't.' 'Why not?' 'I don't have enough film. All my film's allocated for scientific'—I said, 'Bill, you're full of baloney; that is the only picture that anybody will remember from this goddamned flight! None of your volcanoes and craters—Take that picture!' He said, 'No.' So I took the camera and took the goddamned picture. That's the truth of the story. And it's probably on the transcripts too. Did you read it?"

289 *the final, unequivocal verdict: Anders took the picture:* The author broke this news to Borman in March 1988. Susan Borman told her husband he had an apology to make (to Anders). Borman, laughing, said, "I'm not gonna change my story!"

289 *"I've got the same . . . wife I started out with":* Of the twenty-one astronauts who were married when they went to the moon, eleven later divorced or separated from their wives.

293 *When Cernan talks about his moon experiences, not a hint of boredom dulls his words:* Rusty Schweickart says when he first heard Cernan talk about his experiences, at a public appearance in Australia in the mid-1970s, he was astounded. "Gene gave the most absolutely *eloquent, touching, personal* account of the experience of spaceflight and how it has shaped his life—I mean, I sat there next to him *absolutely blown away!* . . . I never thought Gene had that in him. Let alone the willingness to say anything about it."

293 The photograph of Neil Armstrong at Tranquillity Base: There are only a handful of still photographs of Armstrong on the moon; these were taken by Aldrin as part of documentary photography of the lunar module and the moon itself. The photograph in Cernan's office shows Armstrong collecting the contingency sample; it is an enlargement made from 16mm movie film.

297 *"No wonder I was in trouble":* In John Preston, "Buzz: The man who fell to Earth," *The Press* (Christchurch, New Zealand), January 15, 1994.

301 *the best explanation for the origin of the moon:* As described on pp. 353-54 of *To a Rocky Moon* by Don Wilhelms, this theory, proposed by A. G. W. Cameron and William K. Hartmann, accounts for the moon's similarity to the earth's mantle as well as its lack of water and volatile elements, which were driven off by the heat of the impact and escaped into space. While there are still problems to be worked out with the theory, it does appear to explain not only the moon's composition, but the nature of its orbit, which the scientists say would be a direct result of the events that followed the collision.

302 *an answer to the planet's pressing energy needs:* See the author's article, "Shoot for the Moon," *Air & Space/Smithsonian,* December 1991/January 1992, pp. 42-51.

306 *that trip, Mattingly says, is* really *leaving home:* There are leavings, and then there are leavings. Fred Haise says, "I'd like to be the first guy to go to another star, and [watch as] our sun goes away."

A NOTE ABOUT SOURCES

My conversations with the astronauts, totaling hundreds of hours, are the heart of this book, but its framework is built from the extraordinary audio, video, and printed records of their missions. Every word that the astronauts transmitted to mission control was tape-recorded and transcribed. Many of their private conversations were captured by their spacecraft's onboard voice recorder. In the 1980s NASA declassified these tapes, which reveal thoughts and feelings—from irritation to awe—that the astronauts never shared with mission control. When the astronauts were back on earth, they reviewed their flights in detail; tapes and transcripts of these debriefings became another important source. Press kits, technical reports, and handbooks, prepared by NASA and Apollo contractors, provided other key data. Together, these records provided the means to reconstruct the events of the missions in detail, to form a framework on which to display the astronauts' experiences. Many illuminating perspectives came from the astronauts' wives and children, a number of whom participated in the project. Finally, I conducted dozens of background interviews with the people who planned the flights and monitored them from mission control, NASA managers, the scientists who trained the astronauts for their lunar explorations, and astronauts and engineers who worked in support roles before and during the missions.

I also benefited from access to the work of other researchers. At the NASA history office in Houston, David Compton shared many of the interviews he conducted for his own book on Apollo scientific exploration. At the history office of NASA Headquarters in Washington, Lee Saegesser provided me with a number of unpublished sources, including interviews with astronauts and other Apollo participants conducted by writer Robert Sherrod during and after the program. Writer and filmmaker Al Reinart provided a tape of his interview with Apollo 13's Jack Swigert, who had died in 1982. Eric Jones's *Apollo 17 Lunar Surface Journal* (http://www.hq.nasa.gov/alsj/) was an essential source in writing the account of that mission. An excellent source of Apollo photographs is Kipp Teague's web site, Retroweb (www.retroweb.com/apollo).

Another body of source material exists in magazine and newspaper articles published at the time of the missions. Coverage in *Time, Newsweek, Life, National Geographic,* and *The New York Times* alerted me to a number of details to pursue in my conversations with the astronauts. The magazine *Aviation Week & Space Technology* was a helpful source for technical details.

Finally, a number of books deserve special mention. *Chariots for Apollo* by Brooks, Grimwood, and Swenson was an important reference on the Apollo spacecraft development, and many aspects of Apollo mission planning. *Apollo: The Race to the Moon* by Murray and Cox was an excellent source of information on the missions from the point of view of the flight controllers and mission planners. *First on the Moon* by Armstrong, Collins, and Aldrin provided a number of details on the Apollo 11 training, as well as biographical details for the three astronauts. Wilhelms's *To a Rocky Moon: A Geologist's History of Lunar Exploration* was an essential chronicle of Apollo's scientific evolution. And several astronauts' published accounts gave valuable perspectives on their personal experiences before and during their flights, and on the astronaut corps as a whole.

Aldrin, Buzz, and Malcolm McConnell. *Men from Earth.* 2nd ed. New York: Bantam Falcon Books, 1991.

Aldrin, Edwin E., Jr., with Wayne Warga. *Return to Earth.* New York: Random House, 1973.

Anderson, Frank W., Jr. *Orders of Magnitude: A History of NACA and NASA,* 1915-1976. NASA History Series. Washington, D.C.: NASA, 1976.

Armstrong, Neil, Michael Collins, and Edwin E. Aldrin, Jr. *First on the Moon.* Boston: Little, Brown and Company, 1970.

Baker, David. *The History of Manned Spaceflight.* New Cavendish Books, 1981. Reprint. New York: Crown Publishers, 1982.

Beatty, J. Kelly, and Andrew Chaikin, eds. *The New Solar System.* 3d ed. Cambridge, Mass.: Sky Publishing; and Cambridge: Cambridge University Press, 1990.

Borman, Frank, with Robert J. Serling. *Countdown.* New York: Morrow, 1988.

Brooks, Courtney G., James M. Grimwood, and Loyd S. Swenson, Jr. *Chariots for Apollo: A History of Manned Lunar Spacecraft.* NASA SP-4205. Washington, D.C.: Government Printing Office, 1979.

Carpenter, M. Scott, L. Gordon Cooper, Jr., John H. Glenn, Jr., Virgil I. Grissom, Walter M.

Schirra, Jr., Alan B. Shepard, Jr., and Donald K. Slayton. *We Seven.* New York: Simon & Schuster, 1962.
Collins, Michael. *Carrying the Fire: An Astronaut's Journeys.* New York: Farrar, Strauss & Giroux, 1974.
———. *Liftoff: The Story of America's Adventure in Space.* New York: Grove Press, 1988.
Compton, William David. *Living and Working in Space: A History of Skylab.* NASA SP-4208. Washington, D.C.: Government Printing Office.
———. *Where No Man Has Gone Before: A History of the Apollo Lunar Exploration Missions.* NASA SP-4214. Washington, D.C.: Government Printing Office, 1989.
Cooper, Henry S. F., Jr. *Apollo on the Moon.* New York: Dial, 1969.
———. *Moon Rocks.* New York: Dial, 1970.
———. *Thirteen: The Flight that Failed.* New York: Dial, 1973.
Cortright, Edgar M., ed. *Apollo Expeditions to the Moon.* NASA SP-350. Washington, D.C.: Government Printing Office, 1975.
Cunningham, Walter, with Mickey Herkowitz. *The All-American Boys.* New York: Macmillan, 1977.
Duke, Charlie and Dotty. *Moonwalker.* Nashville: Oliver-Nelson Books, 1990.
Gray, Mike. *Angle of Attack.* New York: W. W. Norton & Co., 1992.
Grissom, Betty, and Henry Still. *Starfall.* New York: Thomas Crowell, 1974.
Hacker, Barton C., and James M. Grimwood. *On the Shoulders of Titans: A History of Project Gemini.* NASA History Series. Washington, D.C.: NASA, 1977.
Hawthorne, Douglas B. *Men and Women of Space.* San Diego: Univelt, Inc., 1992.
Heiken, Grant, David Vaniman, and Bevan M. French. *Lunar Sourcebook: A User's Guide to the Moon.* New York: Cambridge University Press, 1991.
Irwin, James B., with William A. Emerson, Jr. *To Rule the Night.* Philadelphia: Holman (Lippincott), 1973.
Irwin, Mary, with Madalene Harris. *The Moon Is Not Enough.* Grand Rapids: The Zondervan Corporation, 1978.
King, Elbert A. *Moon Trip: A Personal Account of the Apollo Program and Its Science.* Houston: University of Houston Press, 1989.
Lay, Beirne, Jr. *Earthbound Astronauts: The Builders of Apollo-Saturn.* Englewood Cliffs, N.J.: Prentice-Hall, 1971.
Levine, Arnold S. *Managing NASA in the Apollo Era.* NASA SP-4102. Washington, D.C.: Government Printing Office, 1982.
Lewis, Richard S. *Appointment on the Moon.* New York: Ballantine, 1969.
———. *The Voyages of Apollo: The Exploration of the Moon.* New York: New York Times Book Company, 1974.
MacKinnon, Douglas, and Joseph Baldanza. *Footprints.* Illustrated by Alan Bean. Washington, D.C.: Acropolis Books, 1989.
Mailer, Norman. *Of a Fire on the Moon.* Boston: Little, Brown, 1970.
Masursky, Harold, G. William Colton, and Farouk El-Baz, eds. *Apollo Over the Moon: A View from Orbit.* NASA SP-362. Washington, D.C.: Government Printing Office, 1978.
Murray, Charles, and Catherine Bly Cox. *Apollo: The Race to the Moon.* New York: Simon & Schuster, 1989.
Mutch, Thomas A. *Geology of the Moon: A Stratigraphic View.* Princeton: Princeton University Press, 1970.
NASA. *Report of the Apollo 204 Review Board to the Administrator, National Aeronautics and Space Administration.* Washington, D.C.: Government Printing Office, April 4, 1967.
———. *The Apollo Spacecraft: A Chronology.* Vols. 1-4. Washington, D.C.: Government Printing Office, 1969-78.
NASA Lyndon B. Johnson Space Center. *Apollo 17 Preliminary Science Report.* NASA SP-330. Washington, D.C.: Government Printing Office, 1973.
———. *Biomedical Results of Apollo.* NASA SP-368. Washington, D.C.: Government Printing Office, 1975.
NASA Manned Spacecraft Center. *Analysis of Apollo 8 Photographs and Visual Observations.* NASA SP-201. Washington, D.C.: Government Printing Office, 1973.
———. *Apollo 11 Preliminary Science Report.* NASA SP-214. Washington, D.C.: Government Printing Office, 1969.
———. *Apollo 12 Preliminary Science Report.* NASA SP-235. Washington, D.C.: Government Printing Office, 1970.
———. *Analysis of Apollo 10 Photographs and Visual Observations.* NASA SP-232. Washington, D.C.: Government Printing Office, 1971a.

———. *Apollo 14 Preliminary Science Report.* NASA SP-272. Washington, D.C.: Government Printing Office, 1971b.
———. *Apollo 15 Preliminary Science Report.* NASA SP-289. Washington, D.C.: Government Printing Office, 1972a.
———. *Apollo 16 Preliminary Science Report.* NASA SP-315. Washington, D.C.: Government Printing Office, 1972b.
Newell, Homer E. *Beyond the Atmosphere: Early Years of Space Science.* NASA SP-4211. Washington, D.C.: Government Printing Office, 1980.
Oberg, James E. *Red Star in Orbit.* New York: Random House, 1981.
O'Leary, Brian. *The Making of an Ex-Astronaut.* Boston: Houghton Mifflin, 1970.
Schirra, Walter M., Jr., with Richard N. Billings. *Schirra's Space.* Boston: Quinlan, 1988.
Surveyor Program [Office]. *Surveyor Program Results.* NASA SP-184. Washington, D.C.: Government Printing Office, 1969.
Taylor, Stuart Ross. *Lunar Science: A Post-Apollo View.* New York: Pergamon, 1975.
Ulrich, George E., Carroll Ann Hodges, and William R. Muehlberger. 1981. *Geology of the Apollo 16 Area, Central Lunar Highlands.* U.S. Geological Survey Professional Paper 1048. Washington, D.C.: Government Printing Office, 1981.
Wells, Helen T., Susan H. Whiteley, and Carrie E. Karegeannes. *Origins of NASA Names.* NASA SP-4402. Washington, D.C.: Government Printing Office, 1976.
Wilford, John N. *We Reach the Moon.* New York: Bantam Books, 1969.
Wilhelms, Don E. *The Geologic History of the Moon.* U.S. Geological Survey Professional Paper 1348. Washington, D.C.: Government Printing Office, 1987.
———. *To a Rocky Moon: A Geologist's History of Lunar Exploration.* Tucson and London: The University of Arizona Press, 1993.
Wolfe, Tom. *The Right Stuff.* New York: Farrar, Strauss & Giroux, 1979.
Worden, Alfred M. *Hello Earth: Greetings from* Endeavour. Los Angeles: Nash, 1974.

ACKNOWLEDGMENTS

Buzz Aldrin, Los Angeles
Judith Allton and Terry Bevel, Lockheed Martin Space Mission Systems & Services, Houston, Tex.
Mary Noel Black, Graham Ryder, and Paul Spudis, Lunar and Planetary Institute, Houston, Tex.
Pete Conrad, Huntington Beach, Calif.
Bob Craddock and Rose Steinat, Smithsonian Institution, National Air and Space Museum, Washington, D.C.
Maureen M. Dilg, National Geographic Society, Washington, D.C.
Jan Evans, Scottsdale, Ariz.
Teresa Fitzgibbon, Albuquerque, N.M.
Mike Gentry and Kathy Strawn, NASA/Media Services, Houston, Tex.
George Hendry and Larry Felieu, NorthropGrumman History Center, Bethpage, N.Y.
Chris Hubbard, The Greenwich Workshop, Inc., Shelton, Conn.
Sherie Jefferson, Irene Jenkins, and David Sharron, Information Dynamics Incorporated, Houston, Tex.
Joe McGregor, U.S. Geological Survey Photographic Library, Denver, Colo.
Carolyn Margolis, Smithsonian Institution, National Museum of Natural History, Washington, D.C.
Edgar Mitchell, Lake Worth, Fla.
Dr. W. R. Muehlberger, Department of Geological Sciences, University of Texas, Austin
Margaret Persinger, Kennedy Space Center, Fla.
Gwen Pittman, NASA Headquarters, Washington, D.C.
Gary Pressel, Adtech Photo Imaging, Houston, Tex.
Christopher Roosa, Falls Church, Va.
Joan Roosa, Gulfport, Miss.
Dr. Leon Silver, Pasadena, Calif.
Kip Teague, Project Apollo Archive, Lynchburg, Va.
Adrienne Wasserman, Astrogeology Team, U.S. Geological Survey, Flagstaff, Ariz.
Robin Witmore, Lick Observatory, Santa Cruz, Calif.

PICTURE CREDITS

The sources for the illustrations in this book appear below. Credits from left to right are separated by semicolons, from top to bottom by dashes.

Slipcase: Front, NASA #AS1714021391. **Back,** NASA; Ralph Morse/Life Magazine © Time Inc.; courtesy Alan Bean—NASA/National Geographic Society Image Collection; NASA #RSC68PC187; NASA #AS12466729, courtesy Andrew Chaikin—NASA #S7035139; Francis Miller/Life Magazine © Time Inc.; NASA #AS12517507. **Spine:** NASA—NASA #S6750531. **Cover:** NASA #AS08142383; plaque created by John Drummond © Time-Life Books, Inc.; NASA #AS1714722526 (inset). **Flap:** Andrew Chaikin.

1: NASA #AS08142383; plaque created by John Drummond © Time-Life Books, Inc.; NASA #AS1714722526 (inset). **2, 3:** plaque created by John Drummond © Time-Life Books, Inc.; NASA #AS08142383. **4:** Marvin Chaikin. **6, 7:** NASA #AS158211166; #AS158211168. **8, 9:** NASA #AS1610917789; #AS16910917790; #AS1610917791; #AS1610917792, panorama assembled by Don Davis. **10, 11:** NASA #AS1714522162; #AS1714522163; #AS1714522165; #AS1714522166, panorama assembled and enhanced by Don Davis. **12:** Astrogeology Team, U.S. Geological Survey, Flagstaff, Ariz. **13:** Detail from NASA #S6818733. **14:** U.S. Geological Survey, Denver. **18, 19:** NASA, courtesy Andrew Chaikin. **22:** NASA #S6925198. **26-29:** Dr. Leon Silver, courtesy Andrew Chaikin. **30:** Courtesy Dr. Farouk El-Baz, Boston University. **32, 33:** NASA #S7020071. **34:** Joe Allen, courtesy Dr. Leon Silver. **35:** Detail from NASA #S6818733. **36:** NASA #S7152276. **40:** NASA #S7131409. **41:** NASA, courtesy Andrew Chaikin (3). **44, 45:** NASA #S7123475. **47:** NASA #S7053300. **48:** Ralph Morse/Life Magazine, © Time Inc. **50:** Itar-Tass/Sovfoto, N.Y. **52, 53:** Ralph Morse/Life Magazine © Time Inc. **54:** NASA #AS158611603. **55:** Detail from NASA #S6818733. **57:** NASA, courtesy Andrew Chaikin. **58, 59:** NASA #AS158711731; #AS158711733; #AS158711735; #AS158711737; #AS158711739; #AS158711740. **60, 61:** NASA #AS158711741; #AS158711743; #AS158711745; #AS158711747; #AS158711749; #AS158711751; #AS158711754. **62:** NASA, courtesy Andrew Chaikin. **64, 65:** NASA #AS158811866. **66:** NASA #AS1858511471. **68, 69:** NASA #AS158511451; #AS158511450. **70, 71:** NASA, courtesy Andrew Chaikin (3); NASA #AS158511435; #AS158511437; #AS158511438. **72:** NASA #AS158711847, courtesy Kipp Teague, Project Apollo Archive. **74, 75:** NASA #AS158511507; #AS158511508; #AS158511509, courtesy Andrew Chaikin; #AS158511511; #AS158511513. **76, 77:** NASA #AS158511514; #AS158511515; #AS158511516; #AS158511517; #AS158511518. **78:** NASA #AS158411324. **80:** AP/Wide World Photos. **82:** NASA #AS15861165. **83:** NASA, courtesy Andrew Chaikin. **84:** NASA #AS159012228. **86:** Robert Nicholson/National Geographic Society Image Collection; NASA. **88:** NASA/National Geographic Society Image Collection. **91:** Astrogeology Team, U.S. Geological Survey, Flagstaff, Ariz. **92:** NASA #S7141138. **93:** NASA, courtesy Andrew Chaikin (3). **94:** NASA/National Geographic Society Image Collection. **96-97:** NASA, courtesy Andrew Chaikin (4). **98:** NASA #AS158811894. **100, 101:** NASA #S7142828. **102-103:** NASA #S7143477; #S7142955. **104:** Berry Berenson/Life Magazine **106, 107:** Dr. Leon Silver courtesy Andrew Chaikin. **108:** Courtesy Jan Evans. **110, 111:** NASA #AS158211120; #AS158211121; #AS158211122. **112, 113:** NASA #AS158211057. **114, 115:** NASA/National Geographic Society Image Collection. **116:** NASA, courtesy Andrew Chaikin (15). **117:** Detail from NASA #S6818733. **119:** Illustration by Don Davis for the Janet Annenberg Hooker Hall of Geology, Gems, and Minerals at the National Museum of Natural History, Smithsonian Institution, Washington, D.C. **120:** Graham Ryder, Lunar and Planetary Institute, Houston, Tex.—NASA #S6945641—Graham Ryder, Lunar and Planetary Institute, Houston, Tex. ; NASA #S7047874. **121:** Graham Ryder, Lunar and Planetary Institute, Houston, Tex. (2)—NASA #S7130341; #S7129179—NASA #S7142955; Graham Ryder, Lunar and Planetary Institute, Houston, Tex. **122:** LOV-125M, courtesy Lunar and Planetary Institute. **123:** Astrogeology Team, U.S. Geological Survey, Flagstaff, Ariz. **124:** NASA #S7157999. **127:** NASA #AS1611818880. **128:** NASA #S7237009. **131:** Courtesy Ruth Goldberg. **132, 133:** NASA #AS1611818895. **134, 135:** NASA, courtesy Andrew Chaikin. **136:** NASA #S7139828. **139:** National Archives Neg. No. 8528 #2. **140, 141:** NASA #AS1611318340; NASA, courtesy Andrew Chaikin (inset). **142:** NASA, courtesy Andrew Chaikin (3). **144, 145:** NASA #AS1610917804; #8028488. **146:** NASA, courtesy Andrew Chaikin (15). **148, 149:** NASA, courtesy Andrew Chaikin. **150:** NASA, courtesy Andrew Chaikin—NASA #S6524899. **152:** William Sievert, Rehoboth Beach, Del. **154, 155:** NASA #AS1611011759; #AS1611011760; #AS1611011761; #AS1611011762. **156, 157:** NASA #AS1611218246; #AS1611218247; #AS1611218256. **160:** NASA, courtesy Andrew Chaikin (3). **162:** NASA. **166:** NASA #AS1610617340. **168, 169:** NASA, courtesy Andrew Chaikin (7). **170, 171:** NASA #AS1611718825. **172-175:** NASA, courtesy Andrew Chaikin. **177:** NASA #AS1611318294. **178-183:** NASA, courtesy Smithsonian Institution Center for Earth and Planetary Studies, National Air and Space Museum, Washington, D.C. **184, 185:** NASA/National Geographic Image Collection. **186:** NASA #AS1714021391. **187:** Detail from NASA #S6818733. **188:** NASA #KSC72P548. **191:** Courtesy of The International Tennis Hall of Fame & Museum, Newport, R.I. **192:** Courtesy Jan Evans. **193:** NASA/National Geographic Society Image Collection. **194, 195:** Ken Heyman-Ronald Diasaro. **196:** Lee Balterman/Life Magazine © Time Inc. **198:** NASA #AS1714822727. **200, 201:** Courtesy Northrop Grumman History Center, Bethpage, N.Y. **202:** Astrogeology Team, U.S. Geological Survey, Flagstaff, Ariz. **203:** NASA #AS1714722465. **204:** NASA, courtesy Andrew Chaikin. **206, 207:** NASA AS1714722526. **208, 209:** NASA #A1714722517; #A1714722519; #A1714722520;

#A1714722495; #A1714722496; #A1714722497. **212, 213:** NASA #AS1713420435. **214, 215:** NASA, courtesy Andrew Chaikin (6). **216:** NASA #AS1713420524—#AS1713420519—#AS1713420530. **218, 219:** NASA #S7255170; NASA #AS1713720979. **223:** NASA #AS1713821039. **224:** AP/Wide World Photos. **226, 227:** NASA, courtesy Andrew Chaikin (6). **228, 229:** NASA #AS1713721011; #AS1713721010; #AS1713721009; #AS1713721008; #AS1713721003; #AS1713721000, all courtesy Lunar and Planetary Institute, Houston, Tex. **230:** NASA #AS1713720990. **231:** NASA, courtesy Andrew Chaikin. **234, 235:** NASA #AS1714021354, courtesy Kipp Teague, Project Apollo Archive. **238, 239:** NASA #AS172444. **240:** Courtesy W. R. Muehlberger, University of Texas, Austin. **244:** NASA/National Geographic Society Image Collection. **246:** NASA, courtesy Andrew Chaikin (3)—NASA #AS1714622294. **248, 249:** NASA, courtesy Andrew Chaikin. **250, 251:** NASA/National Geographic Society Image Collection. **252:** NASA #AS1713420481—#S7255417. **255:** NASA, courtesy Andrew Chaikin (4). **256:** NASA #AS1714922859. **258, 259:** NASA #AS1716224053. **260:** NASA #AS1715223274. **262:** NASA #AS1715223308. **264:** NASA #AS1715223401. **266, 267:** NASA #7255837; courtesy Jan Evans. **269:** NASA #73HC181. **270, 271:** NASA #AS1713620695. **272, 273:** NASA/National Geographic Society Image Collection. **274, 275:** NASA #AS1713420421; #AS1713420422; #AS1713420423; AS1713420424; #AS1713420425; #AS1713420426; #AS1713420427, panorama assembled by Don Davis. **276, 277:** NASA #AS1713720910. **278:** Detail from NASA #S6818733. **279:** Courtesy Pete Conrad. **281:** © Joyce Brock, 1998, courtesy Edgar Mitchell. **282:** NASA #S8942879. **285:** © Phil Schofield, 1994. **287:** © Douglas Merriam. **290:** © Helen Morse, courtesy Andrew Chaikin. **292:** © David Nance, Photographer. **294:** NASA, courtesy Andrew Chaikin. **297:** Photo by Lazlo, courtesy Buzz Aldrin. **298:** NASA #S091280858. **304:** Courtesy Joan Roosa. **305:** NASA #8232925. **307:** © David Nance, Photographer. **311:** UCO/Lick Observatory photo, Santa Cruz, Calif. **312:** NASA #S6931743—#S6853187—#S6931741—#S6938859—#S6853187. **313:** NASA #S7250438—#S6931742—#S6938866—#S7216660—#S7250438. **314:** NASA #S6938862—#S6962224—#S7156478—#S6962224 #S7216660. **315:** NASA #S7055635 (2)—#S7250438—#S7137963—#S7055635. **316:** NASA #S6842906—#S7152266—#S7152280—#S7216660. **318:** NASA #S6826668—#S6851093—#S6918569. **319:** NASA #S6931959—#69HC498—#S6952336—#S6960662—#S7017851—#S7130463. **320:** NASA #S7156246—#S7249079.

INDEX

Numerals in italics indicate an illustration of the subject mentioned.

N

O

P

Q

R

S

T

V

W

Y

Z

ANDREW CHAIKIN, A MAN ON THE MOON
VOLUME III: LUNAR EXPLORERS

Copyright © 1994 by Andrew Chaikin
All rights reserved.

Grateful acknowledgment is made for permission to reprint "Spacewalk" and "Quietly, Like a Night Bird" from *Hello Earth: Greetings from Endeavour* by Alfred M. Worden (Nash Publishing Company, 1974). Reprinted by permission of the author.

First published in 1994 by Viking Penguin, a division of Penguin Books U.S.A. Inc., as Book 3 of *A Man on the Moon: The Voyages of the Apollo Astronauts* by Andrew Chaikin.

This Time-Life edition is published by arrangement with Viking Penguin, a division of Penguin Putnam Inc.

Cover, text design, and captions © 1999 Time Life Inc.

All rights reserved. No part of this book may be reproduced in any form or by any electronic or mechanical means, including information storage and retrieval devices or systems, without prior written permission from the publisher, except that brief passages may be quoted for reviews.

Second printing 1999. Printed in U.S.A.

School and library distribution by Time-Life Education, P.O. Box 85026, Richmond, Virginia 23285-5026

TIME-LIFE is a trademark of Time Warner Inc. and affiliated companies.

Library of Congress Cataloging-in-Publication Data
Chaikin, Andrew, 1956-
A man on the moon / by Andrew Chaikin and the editors of Time-Life Books.
p. cm.
Includes index.
Contents: 1. One giant leap — 2. The odyssey continues — 3. Lunar explorers.
ISBN 0-7835-5675-6 (v. 1).—ISBN 0-7835-5676-4 (v. 2).
—ISBN 0-7835-5677-2 (v. 3).
1. Project Apollo (U.S.)—History. 2. Space flight to the moon—History. I. Title.
TL789.8.U6A5244 1999
629.45'4'0973—dc21 99-15449
CIP

TIME® LIFE BOOKS

Time-Life Books is a division of Time Life Inc.

TIME LIFE INC.
PRESIDENT and CEO: George Artandi

TIME-LIFE BOOKS
PUBLISHER/MANAGING EDITOR: Neil Kagan
SENIOR VICE PRESIDENT, MARKETING: Joseph A. Kuna
VICE PRESIDENT, NEW PRODUCT DEVELOPMENT: Amy Golden

EDITOR: Lee Hassig
DIRECTORS, NEW PRODUCT DEVELOPMENT: Mary Ann Donaghy, Elizabeth D. Ward, Paula York-Soderlund

Design Director: Cynthia T. Richardson
Assistant Art Director: Janet Dell Russell Johnson
Senior Marketing Manager: Paul Fontaine
Project Manager: Karen Ingebretsen
Associate Editor/Research and Writing: Ruth Goldberg
Senior Copyeditor: Judith Klein
Page Makeup Coordinator: Kimberly A. Grandcolas
Editorial Associate: Patricia D. Whiteford

Special Contributors: Andrew Chaikin (captions); Marilyn Murphy Terrell (research); Christine Stephenson (text); John Drummond, Belen Price, Don Davis, Totally, Inc. (mosaics); Mary Gasperetti, Jenifer Gearhart (design and production); Marianne Dyson, Amanda Stowe (picture research); Antheus L. Bowden (picture coordination); Sunday Oliver (index).

Correspondents: Maria Vincenza Aloisi (Paris), Christine Hinze (London), Christina Lieberman (New York).

Director of Finance: Christopher Hearing
Director of Book Production: Patricia Pascale
Director of Imaging: Marjann Caldwell
Director of Publishing Technology: Betsi McGrath
Director of Photography and Research: John Conrad Weiser
Director of Editorial Administration: Barbara Levitt
Manager, Technical Services: Anne Topp
Page Makeup Manager: Debby Tait
Senior Production Manager: Ken Sabol
Production Manager: Virginia Reardon
Quality Assurance Manager: James King
Chief Librarian: Louise Forstall

Separations by the Time-Life Imaging Department

OTHER PUBLICATIONS

COOKING
Weight Watchers® Smart Choice Recipe Collection
Great Taste~Low Fat
Williams-Sonoma Kitchen Library

DO IT YOURSELF
Total Golf
How to Fix It
The Time-Life Complete Gardener
Home Repair and Improvement
The Art of Woodworking

HISTORY
Our American Century
World War II
What Life Was Like
The American Story
Voices of the Civil War
The American Indians
Lost Civilizations
Mysteries of the Unknown
Time Frame
The Civil War
Cultural Atlas

TIME-LIFE KIDS
Student Library
Library of First Questions and Answers
A Child's First Library of Learning
I Love Math
Nature Company Discoveries
Understanding Science & Nature

SCIENCE/NATURE
Voyage Through the Universe

For information on and a full description of any of the Time-Life Books series listed above, please call 1-800-621-7026 or write:

Reader Information
Time-Life Customer Service
P.O. Box C-32068
Richmond, Virginia 23261-2068

Lunar Explorers is the third volume of the three-book set
A MAN ON THE MOON.
The other titles are:
One Giant Leap
The Odyssey Continues